The Reason for Science

Donald R. Strombeck

Stonegate Publishing, Davis, California

Stonegate Publishing
633 Marina Circle
Davis, California 95616
ISBN 0-9619272-0-8
Printed in the United States of America

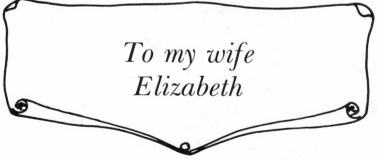

To my wife
Elizabeth

Contents

Preface And Acknowledgments

Can science, which claims to produce pure knowledge, "value-free," be said to have a purpose? People have pursued science purposefully for hundreds of years. Most people believe the reason for science is to benefit humanity. The United States Government is preparing to spend one to two billion dollars to map the human genome and four to five billion to build a Superconducting Super Collider for investigating subatomic structure. How mapping the human genome will help us is not at all clear. Yet we are being convinced that such an accomplishment holds major importance for humanity's future. The importance of better understanding subatomic structure is equally uncertain. Yet a huge project goes forward because of faith that knowledge is our destiny.

The human benefits from the major project in biological science and the one in atomic physics are thus both in doubt. In addition, a major conflict disturbs these two branches of science. Biological fact and theory conflict with philosophical interpretations stemming from physics. Moreover, modern physics and evolutionary biology each claim to provide the sole paradigm for modern thought. The new physics is claimed by some to *be* philosophy, maybe even *the* philosophy of the future. For others physics is no longer the yardstick of science, particularly in studies of human beings. In all of recent science, what is really new? How has the new contributed to human progress? We continue to ask who we are and what kind of world we live in. For finding such answers, what paradigm should we follow? Do we understand the reason for science?

I am deeply grateful for the editorial assistance contributed by Walter R. Hearn who made numerous constructive suggestions. I did not always follow his advice and am solely responsible for the book's final content. I am also indebted to my wife Elizabeth for many hours of typing, proofing, and editing. She also sustained me with immeasurable encouragement and belief in the work. Lincoln Hurst has been invaluable as a good friend and teacher; he has contributed in many unseen ways. I am indebted to Marlene Ekman for the artwork on page 137.

Introduction

What Is Science?

Science is generally defined as knowledge. Knowledge, however, includes everything that we perceive and grasp by our minds, so science cannot be the only source of knowledge. Plato distinguished knowledge from mere belief by limiting the scope of knowledge to science. Today science is accepted as knowledge only when based on the experiences of human sensory perceptions; scientists reject (as science) anything based only on intuition or belief. A belief that can be "justified as true," however, does become knowledge. Scientists are engaged in justifying and creating truth for beliefs.

Truth is generally defined as a verified fact, an accurate statement or proposition, an established principle, or a fixed law. A truth is established by its objectivity and universality; the conditions of its definition must conform to reality and be accepted by most human beings. Truths are established by correspondence (by deciding how a belief compares with facts), and by coherence (by deciding how a belief fits in with other strongly held beliefs).

All human beings look to sources other than science for knowledge, sources which do not contribute to the experiences of human sensory perceptions. They are sources of "extrasensory" perceptions, where knowledge is gained by intuition or inspiration. The basis for such innate or *a priori* knowledge is metaphysical, beyond scientific understanding. Some individuals may credit the source to a design or to God; others may consider it inherent in nature, having no explanation other than as a manifestation of an evolutionary progression. Along with the scientists' justified true beliefs, these empirically unproven forms of knowledge contribute to what human beings hold to be true.

Knowledge enables people to describe the world and be able to evaluate and prescribe for improving their condition. Moreover they seem to have believed almost universally that with sufficient knowledge they could understand the universe and the essence of their brief biological existence. It is hardly surprising that science has appeared in some form in the recorded endeavors of every human culture, making its contribution to knowledge important for all societies. Science has frequently been pursued merely for acquiring knowledge; such a pursuit is sometimes described as "pure science." Science has also been pursued with some practical application in mind, leading to the development of technology. On an overall basis, people pursue science to improve understanding of explanatory "truths" about the world, and about what is of value.

Understanding means knowing or grasping the meaning, importance, intention or motive of something. For some things, such comprehension may require considerable interpretation in order to perceive clearly the thing's true nature. Understanding is necessary for accepting any item of knowledge as a reliable fact or truth.

Value is a quality of something according to which it is thought of as being more or less desirable, useful, estimable, and important; it is a quality defining the degree of worth of something.

Possession of knowledge and understanding does not necessarily give people an ability to find truth. Only some persons are said to be wise, to have the power of discerning and judging correctly, of discriminating between what is true and what is false and between what is proper and improper. *Wisdom* enables some to find truth and what is proper by making the best use of knowledge, experience, understanding, and innate attributes such as insight and inspiration. People today continue to seek knowledge, believing that it will greatly enhance their understanding and wisdom. What better place to seek knowledge than in science? The quest for scientific knowledge actually began early in recorded history.

Chapter 1

In the Beginning

Cosmology: Expression of Knowledge

Perhaps the earliest human knowledge is reflected in cosmology. A *cosmology* is a construction of theories or philosophies concerning fundamental principles underlying the universe. As perceptions on the nature of the universe have accumulated, people have used that knowledge to develop cosmologies. Cosmologic beliefs usually pertain to (a) the nature of events during the creation of the universe, and (b) the existence or nonexistence of gods exercising some degree of control over the universe's operation.

Cosmologies from before the time of the earliest Hebrews attribute the creative forces driving the world to more than one god. They depict the universe as created out of chaos, but not as a creation of structures either from nothing or from totally disordered matter. For example, the Mesopotamian creation story describes Marduk, the Babylonian champion of gods, as crushing the powers of chaos led by the goddess Tiamat.[1] A creative force thus prevailed as the outcome of a battle between gods in which the forces of chaos were defeated. With such "story myths" the creation of the world could be explained on the basis of personal knowledge of the good and evil of human nature. Imagining that the gods must be like themselves, early thinkers attempted to answer questions about the world by attributing human drives and emotions to the gods. Such imaginative cosmologies served human beings until they were able to accept something more.

The oldest complete cosmology giving a single divinity responsibility for the creation and ordering of the universe is found in the Hebrew Bible. Its first book, the book of Genesis, begins with a description of God's acts to create an ordered, determined universe. Order emerged out of chaos, not instantaneously, but as an exacting process of change. Chaos was not a conflict between forces of good and evil; rather, it was a void, a nothing. Born into cultures that taught them to believe in human-like gods, the earlier peoples had been unable to formulate such an abstract concept as order being formed from a void. Too sophisticated for human beings to imagine, the construction of the world from a void was a brilliant concept that is the basis for current science's cosmology.

According to Genesis, life was created in a stepwise process, beginning with simple forms and becoming increasingly complex until finally human beings were formed, God's ultimate creation. Charles Darwin saw the forms of life as in a hierarchy, the history of which could be documented by discovering constituent taxa or groups which appear to stand in relation to one another as ancestors. A rudimentary form of that concept is present in the biblical account of creation.[2] God's first revelation suggests a design that can operate, at least in part, through evolutionary processes; the seeds of Darwinian evolution may have been sown in the first chapter of Genesis.

Viewing the first Genesis creation narrative as consistent with an evolutionary process, however, does not imply that any or all orders of life have evolved from a preexisting, less complex order. Long before scientific observations led to theorizing about evolutionary processes for the world, God may have suggested such a process to early peoples because it would not greatly upset their understanding of the universe. God began his revelation to human beings at the level of knowledge and understanding they had at the time. If ancient cultures believed in some rudimentary form of evolution, God may have been inclined to allow his people to maintain some remnant of that concept rather that to totally reject it. Examples of evolutionary beliefs are definitely found among the early Greeks: in the sixth century B.C., Anaximander thought that humans derived from fish of the sea; perhaps two centuries later Empedocles held primitive views of biological development based on haphazard combinations selected by a kind of "survival of the fittest."[3]

The second Genesis creation story does not describe the organization of an ordered and determined universe created out of disorder. It does not deny the message of the first Genesis creation account but rather complements it. The focus of the second narrative is the human person, depicted as a fully developed biological entity. Taken literally, formation of the first man from the dust of the earth within a period of one day would seem to imply that he was created by an instant transformation. Creative acts requiring such a miraculous genesis were commonly accepted by early peoples; today they are unacceptable for many and hence considered to be merely ancient myths. When we interpret the Bible or any other ancient text, problems arise:

> . . . because the language and thought forms we are studying are not in themselves intelligible without interpretation, but our own language and thought forms are not adapted to fit them, therefore interpretation is always problematic and accompanied by distortion.[4]

The narrative as written does not preclude a slow evolutionary process for human biological development.

The intent of the whole book of Genesis is not to be

merely a cosmology, a historical account of the time-course of important events since the creation of the world. If it were, some apparent discrepancies in the two creation narratives and in other stories would have been reconciled. Viewing them as separate revelations with different purposes, the early Hebrews wisely included both narratives in their Scriptures despite certain inconsistencies. Speaking to his people at different times over a span of many centuries, God revealed his presence and purpose in a manner they could understand. As a result, the revelations at the time of Moses could very well have seemed primitive to the intelligent Hebrew priests who received revelations about creation some six to eight centuries later.

As God's holy ones, those priests were responsible for protecting and maintaining the older traditions, even though their inclusion might make the Hebrew Bible appear to contain a series of incongruities. The cosmology of Genesis changed from one version to the next through a process of progressive revelation. In one creation story God was speaking to primitive people; in the second story, to sophisticated Hebrew priests. Increasing human knowledge, reason, and imagination are not credited for the changes in understanding God; for that understanding the Hebrews were indebted to God's progressive revelation.

When the first human being was created "out of the dust of the earth," nothing in particular distinguished the new creature as unique. The human creature would have differed only slightly from other animals whose behavior for ensuring individual survival and perpetuation of the species was solely under control of physiological forces. As with subhuman animals, nothing inherent in this "creature of the dust" enabled it to interact with God, its creator. God's supreme act was to give this particular bit of creation a *spirit* (or "soul"); that creative act was accomplished by "breathing into his nostrils the breath of life,"[5] thereby imparting a potentiality unique to human beings. That act distinguished humans from every other form of life; it gave them the potential to be the special creatures they were designed by God to be. By that unique gift, human beings became spiritual "living creatures," no longer constrained to the merely biological existence of all other living forms.

The First Freedom of Human Choice; Knowledge or Immortality?

God gave human beings a freedom of choice not known to other forms of life, freeing them from determined behavior dictated entirely by physiological needs. For example, our first human ancestors did not have to relate to each other by sheer instinct, driven for procreation, satisfaction of hunger, or maintenance of a territory. Freedom of choice as part of their human potentiality enabled them to be party to a conditional covenant with God, something impossible for mere animals. The token of human obedience to God exacted in that covenant shows that the "spiritual life" breathed into them indeed made them more than "beasts of the field."

The very first human pair, equipped with a unique potentiality by the "breath of life," lived free from any needs or concerns—"in paradise." Only one condition was placed on their existence there: "of the tree of the knowledge of good and evil you shall not eat, for in the day that you eat of it you shall die."[6] Given freedom of choice as part of their potentiality, the first lesson in being human was to learn to live responsibly under an authority greater than themselves.

The "knowledge of good and evil" should not be interpreted literally but rather be understood as meaning "knowledge of everything."[7] Hebrew idioms frequently employ opposites to convey the idea of totality. For example, the phrase "we cannot speak to you evil or good" means that we have nothing to say. The first humans were warned by God that the totality of knowledge was forbidden to them, and that disobedience would result in loss of their paradise. Humanity was being warned by that one statement that we cannot have access to God's knowledge and power, and that we are subject to God's authority.

Several alternative explanations of "knowledge of good and evil" are based on literal interpretations in which the creation story is compared with and found similar to traditions in cultures contemporaneous with the writing of Genesis.[8] One interpretation holds that "knowledge of good and evil" meant knowledge of the spells and incantations by which people have often sought to control friendly and hostile spiritual forces. That interpretation has little support, however: occult practices are expressly forbidden in many Old Testament passages.[9]

Another interpretation held by some theologians suggests that the phrase referred to a wide range of experiences alien to the Hebrew culture. For example, the forbidden fruit has been suggested to be symbolic of the foreign gods and fertility cults that were a recurring temptation to the Hebrews from the time they settled in Canaan. That interpretation gives insufficient credit to the Hebrew storyteller narrating God's revelation.

The most reliable interpretation of "knowledge of good and evil" will be based on an honest evaluation of what the author wrote and not on its similarities to what is found in other traditions at the time of writing. This revelation was of a unique, abstract concept that was framed to be understood for Hebrews in their cultural infancy. The writer used terms understood by his own people and a background of their own religious beliefs for recording God's revelation. The revelation, coming at a time when people believed in many anthropomorphic gods, had to deal with that situation. (God's revelation in the Bible is no more anthropomorphic than is Plato; both recognize the total mystery of God and use anthropomorphism to try to comprehend it.[10]

A valid interpretation cannot be based on what the author *should* have thought or said in the context of twentieth-century ideas. Biblical portrayals are sometimes offensive to modern human reason. If the same revelations were newly presented today, they might be expressed in a structure based on current understandings of Darwinian evolution and quantum physics, a more plausible structure for at least some people. The language of God's revelation, whether expressed to individuals thirty centuries ago or today, must be an accommodation to the limits of the human capacity at some particular time. As has been pointed out, sound interpretation that does justice to the early Hebrew writers, and to their idiomatic expressions of totality, is that the "knowledge of good and evil" meant a knowledge of everything.

The first human choice was thus between paradise, with the promise of eternal life, and the temptation to seek the forbidden total knowledge. That the ultimate human desire is to attain knowledge above all other needs or wants is a unique feature of the biblical creation account. Knowledge was not of such paramount importance for any other culture or religion of antiquity. Rather than seeking the total knowledge possessed by a god, other cultures such as in Mesopotamia tended to focus on the inevitability of death and hence on a quest for immortality.[11] Today, in our own culture, people do seem to make total knowledge their goal; they seek to answer all possible questions of the natural world.

In the Genesis account of the first humans, immortality was of no concern because they were given the opportunity of eternal life; no divine prohibition kept them from eating fruit of "the tree of life." The idea that humanity originally *had* immortality and then lost it was a unique concept at the time of the early Hebrews. In *Understanding Genesis*, Nahum Sarna has this to say:

> By relegating the tree of life to an insignificant subordinate role in the Garden of Eden story the Bible dissociates itself completely from this preoccupation. Its concern is with the issues of living rather than with the question of death, with morality rather than mortality. Its problem is not the mythical pursuit of eternity, but the actual relationships between man and God, the tension between the plans of God and the free-will of man. Not magic, it proclaims, but human action is the key to a meaningful life.[12]

People have probably always wondered about their fate after death. Some early cultures may have been preoccupied with a day-to-day struggle for survival but others sought immortality through magic and ritual. Preoccupation with death developed into one of the most important human activities once ancient civilizations had solved the basic problems of obtaining food and shelter. Meticulous efforts to preserve the dead were made during Egyptian and other ancient dynasties. Some of the monuments built for entombing the dead rank as wonders of the ancient world. In the now-outlawed Hindu practice of suttee, a wife, regarded as her husband's possession, was killed at his death in order to remain the property of the dead man and accompany him into the unknown of death. Human endeavors to continue life beyond death have thus ranged from the magnificent to the barbaric.

Yet despite so much searching, the question remained: does any form of human existence persist after death of the body? In the Genesis account, the "tree of life" represented the human quest for immortality.

Why Choose Knowledge?

After the first humans were told, "but of the tree of knowledge of good and evil you shall not eat, for in the

day you eat of it you shall die," the tempter told the
woman, "you will not die."[13] He continued, "For God
knows that when you eat of it your eyes will be opened,
and you will be like God, knowing good and evil."[14] The
first expression of human free will (the ability to choose
between two options) was unconstrained—but not without
consequences. This simple creation narrative describes
what God wanted the first humans to know about him; it
contains God's first revelation, or dialogue, to humanity. A
covenant then develops; God grants the chance for immortality, and human beings have an absolute requirement of obedience. Created innocent, with freedom of
choice, they could choose to be obedient or could deny
their obligation. Before committing the error of disobedience, Adam and Eve did not appreciate that they had no
need for total knowledge, which only God possessed.

With the tree of life seen as a symbol of the greatest
human desire throughout the ancient world, how does one
explain the unique appearance of the tree of knowledge as
a more important goal than eternal life? Seeking knowledge is not important to technologically primitive people.
Such people are content to live with their myths, letting
folklore provide explanations of the world and their life in
it. They do not strive to achieve the total knowledge
sought by the first man and woman, or by modern persons
who believe that acquisition of knowledge will solve all
problems. The first humans may have chosen to seek total
knowledge because they did not believe they would lose
immortality; they were assured of their immortality by the
serpent. They chose, that day, whom humanity would
follow.

The first humans in their fallen state (after disobeying
God) did not gain total knowledge from the tree of
knowledge of good and evil; their continued ignorance,
plus new burdens of suffering and death, made them even
more dependent. Laws and revelations from God became

humanity's only means to discern "good and evil" and
survive in this world.

Revelation or Sublime Mythology?

Myths develop to answer deep questions about life,
society, and the world. Should the biblical account be
considered a myth that merely perpetuates a debasing of
endeavors for improving human existence? In *The Myths of
Israel* Amos Fiske wrote:

> But why was the eating of the fruit forbidden? Is there not
> in this an expression of sadness that man should persist in
> knowing and doing for himself, and so incur the penalties
> of achievement, and of a divine solicitude that he might
> remain innocent and happy and forever a child? Once
> tasting of knowledge and coming to the choice of good and
> evil, he is debarred from the tree that would furnish the
> antidote to the bitter penalty.[15]

Fiske saw sadness in humanity's being saddled with the
bitter penalty of death for achievement gained by knowing
and doing solely according to human inclinations.

The biblical accounts of creation and fall are unique and
brilliant because they show such profound insight into the
human situation. The accounts reveal as inherent to
humanity a quest for an ultimate attribute possessed by
neither innocent nor fallen humans: "the knowledge of
good and evil," a human obsession to possess total knowledge, over innocence, freedom from suffering, or eternal
life.

The Promise of Knowledge

Human beings have steadfastly persisted in knowing
and doing for themselves, believing that knowledge holds

the antidote to the bitter penalty. That belief has changed little since the first documentation of the importance of knowledge in the book of Genesis. People continue to seek knowledge for answers about their identity, reasons for existence, and ultimate fate. Today many people believe that with sufficient knowledge, society can be constructed anew.[16] In fact, knowledge—or more specifically, science—has been largely credited for the earliest advancements by our primitive ancestors.[17]

Continuation of scientific activity is said to promise what was denied in the Garden of Eden:

> Nevertheless, modern science is guided overall by this moral political intention: a lifting up of downtrodden humanity; a reversal of the curses laid upon Adam and Eve; and ultimately, a restoration of the tree of life, by means of the tree of knowledge.[18]

Future achievements in science are believed by many to promise greater understanding of human motives and behavior, so that eventually all human problems can be solved. Moreover, scientific knowledge is generally viewed as essential for continuing human existence:

> And yet, fifty years from now, if an understanding of man's origins, his evolution, his history, his progress is not the commonplace of the schoolbooks, we shall not exist . . . . knowledge is our destiny . . . . The personal commitment of a man to his skill, the intellectual commitment and the emotional commitment working together as one, has made the ascent of man.[19]

To appreciate why human beings began to seek knowledge and have continued through the present to pursue its special scientific form, we must consider why people believe the fruits of scientific endeavors to be useful. In what follows we will examine human efforts to pursue science for the achievement of knowledge, and the view that such efforts have indeed made possible the "ascent of man."

Chapter 2
Science in the Beginning

Seeking and Finding Order

Since earliest recorded history people have lived with a conscious or unconscious awareness of *order* in the universe. Order is obvious in the regular sequence of day and night and in the predictable cycling of seasons year after year. For the most part, people have believed the order to be created, a consequence of design by single or multiple supernatural beings. Events in the universe were thus determinate or at least appeared so because each one was specified in a divine plan for the present and future world. In an ordered, determinate universe where everything ran mechanically, like a clockwork, nothing could happen randomly or spontaneously. The view held by scientists that the universe is comprehensible would be difficult in a random, chaotic, spontaneous universe.

Many imagine that knowledge of the universe's design should offer insights into the meaning of human existence. An innate human quest for total knowledge, revealed in Genesis, has for many people superseded all other desires. Such a quest could be approached from two totally different directions.

On the one hand, people could revere the creator of the universe, expecting whatever understanding they needed or desired to be provided by the divine designer through gifts of revelation. For centuries such an understanding was provided by a "salvation history" found in the Holy Bible of Christians and Jews. For many the Bible still offers the only true answers to the most basic human questions.

On the other hand, could it be possible for people themselves to develop a way of thinking to lead to an understanding of the unknowable? Any lack of confidence in divine concern for human welfare might seem to justify resuming the quest that faltered in the Garden of Eden.

The two options have thus been to (a)rely on God and on God's design, which includes a concern for human welfare, or (b)embark on a quest for purely human knowledge, to gain understanding and perhaps control of human destiny.

From Understanding Order to Gaining Control

It is only recently that control by humans over their destiny has seemed possible. Before the last few centuries people looked to magic or religion in a rather futile hope of exercising some control over the world in which they lived. Yet they continued to pursue the goal of acquiring knowledge in the hope of gaining an understanding of life's meaning.

Now, however, we have discovered that our environment can be controlled and manipulated to considerable advantage. Our remarkable technological accomplishments have led many to believe that the possibility of improving all aspects of the human condition is within our grasp. Moreover, we tend to credit scientific endeavors for developing the present human condition. For some, with even a stronger faith in science's potential:

> Science is inherent in the oldest parts of human achievement, and in man himself. There would never have been

any men, if their sublime ancestors had not turned themselves into men by pursuing what, in essence, was primitive scientific activity: the mastery of themselves and their environment.[1]

Such a view congratulates humanity for molding and developing an innate ability possessed by subhuman primates and for pulling itself up by that process to rise above the limited potential of animals. If primitive scientific activity really was responsible for the evolution of human beings from primates, it would be reasonable to assume that with continuing scientific achievement over sufficient time, humans with unlimited powers would eventually achieve everything toward which their efforts are directed.

Progress in Science and Technology Is Discontinuous

Throughout history, the progressive nature of technological development has been intermittent. Technology and science in general have experienced periods of enlightenment in which predictability and control have made phenomenal advances, but there also has been centuries during which little new was reported. Consider the flourish of scientific achievement in the early Greek civilization, then fifteen centuries of very little progress, followed by a scientific revolution in Europe seventeen to twenty centuries after the Greeks. How does one explain such periodicity? Why was there not a similar revolution elsewhere in the world? In China the basic prerequisites for the birth of a scientific revolution had been in place centuries before the European scientific revolution occurred. Why didn't it happen in China? Some understanding of the history of science is necessary before considering these questions.

Scientific Progress: New Insights on Old Questions

Most people have great hopes for what science can achieve, because they believe that science is cumulative.

But scientific knowledge is not automatically cumulative, in the sense that attaining a minimum critical mass of information inevitably produces new forms of technology or revolutionary theories. Nevertheless it is unusual for anything new in science to be developed except on the basis of what is already known. The ancient Greeks may appear not to have needed any earlier scientific achievements in order to make extraordinary observations and formulate remarkable conclusions about events consistently seen in the universe. Yet records from antiquity, though meager, indicate that certain scientific achievements of the early Greeks were facilitated by observations recorded at earlier times.[2] For example, in the sixth century B.C., Thales of Miletus predicted an eclipse of the sun, possibly by using Babylonian records of such events.[3] Such information was not available to Thales from Greek astronomers, since the Greek civilization had not been in existence long enough to accumulate sufficient knowledge on the movement of celestial bodies.

The Greek scientific achievements are generally viewed as unique and remarkable. But in some cases it may be that the contributions by the Greeks have been greatly overstated. According to Toulmin, the forecasting procedures developed by Kidinnu and other calendrists of classical Babylon were technically in advance of any that the Athenians ever invented for themselves:

> . . . in this respect, the computations employed today at the United States Naval Observatory, in preparing tables of astronomical ephemerides, represent refinements of procedures first worked out by Kidinnu and his Babylonian colleagues, rather than of anything devised independently by Eudoxus or Heraclides or Aristotle.[4]

Was the advancement of science by the early Greeks a kind of scientific revolution? A significant feature of scientific revolutions seems to be finding a set of new ways to examine old ideas.[5] The ancient Greeks found new ways of looking at old questions about the nature and organization of substance and life (that is, the microworld) and the organization and evolution of the universe (the macroworld). Interest in those areas continues as an important part of current science. Science is still examining old questions in new ways.

An Atomic Theory (for All Ages?)

Atomic theory, the concept that matter is made up of constituent particles too small to be seen, in perpetual motion in a void, with each particle rigid, solid, and indivisible, was defined by Leucippus and developed further by Democritus in the fifth century B.C.[6]

Little new was introduced to improve on that early exposition on the nature of the microworld until the early nineteenth century, when the chemist John Dalton used the Greek ideas on atomic theory to explain his observations on the constant proportions in which chemical substances combine.[6] Dalton's theoretical explanations were ingenious descriptions of the quantitative laws of chemistry. Democritus's atomic theory must be considered even more ingenious.

Without any experimental evidence either to suggest or to support his ideas, what Democritus proposed was pure speculation. His theory on the nature of matter was determinate; everything was the result of cause and effect among the atoms, the motions of which were rigidly governed by laws. If "chance" had any role in the interaction of atoms it was not to be found in the sequence of cause and effect but would reveal itself in one's inability to predict the outcome. That inability became the basis of present day indeterminism.

Atomic theory was not an accidental discovery in early Greece. Its formulation represented Democritus' answer to a century-and-a-half-old question: What is the composition or substance of all matter, and how can a knowledge of that substance account for the changes seen in the world?

The modern quest for unifying theories on subatomic particles and the interacting forces determining their properties is not unique. The ancient Greeks also sought unifying explanations; it was their idea that the quantity of matter remains always the same. The concept that matter is never being created or destroyed but merely changed from one form to another was scientific dogma well into the twentieth century.

Seeking a unifying explanation, Heraclitus believed that "all things are one"; for him that one thing was fire.[7] His "grand unifying theory" was that everything is a form of fire, either in fire's progressive condensation through a series of steps to form gas or moisture, liquid water, and subsequently earth, or in the formation of fire through a reversal of that process. His theory provided a framework for the universe's beginning as fire or energy, which then evolved into material structures. That simplistic hypothesis, which must have been based on virtually no empirical scientific evidence other than common observations, is philosophically no different from the current "big bang" theoretical explanation for the creation of the universe where the most elementary particles of matter arose solely from energy. It is remarkable that the concept of all matter being a form of "fire" remained essentially untested until the nineteenth century. Although much recent empirical evidence has provided knowledge of the chemical composition of matter, current concepts have changed little.

Current cosmology also proposes that creation sprang from a void, that matter was created out of nothing. That "wild" philosophical idea, objectionable to common sense, is also found in the biblical creation story—but not in the thought of Heraclitus. He believed that the world "ever was, and is, and shall be" ever-living fire and forms thereof. In his scheme there could have been no creation from a void.

Current ideas on the composition and origin of matter are essentially the result of rethinking an old question. The new way of thinking that resulted was the atomic theory. The twentieth-century understanding of matter describes the structure of atoms on the basis of empirical evidence and theoretical considerations—after centuries of increasingly sophisticated experimental techniques and mathematical explanations. Why the early Greeks were led to develop an ingenious atomic theory, which lasted more than twenty centuries, remains an unanswered question. Why that theory was not improved on for over twenty centuries is also an interesting question.

Search for Order in the Cosmos

The design of the universe was held by the ancients to be based on beauty, simplicity, symmetry, and perfection. The circle was considered a symbol of perfection; therefore, the design of the universe had to have planets moving in circular orbits. With the earth believed by philosophers to be the center of the universe, it was feared that its removal from that position would result in disorder or chaos, including a breakdown of all morals.[8]

About 260 B.C., Aristarchus, possibly building on earlier Greek views of the universe, developed a heliocentric theory in which the earth moved in an orbit around the sun.[9] Aristarchus perceived that to an observer on the earth in orbit, the relative position of the stars should seem to change. Aristarchus realized that he could not prove such changes because they were too small to be measured.

That proof was not gained until development of the telescope permitted more sophisticated techniques of observation.

The heliocentric theory, a most amazing achievement of the early Greeks, was revived in the early sixteenth century by Nicolaus Copernicus. The name Aristarchus was written by Copernicus in the margin of one of his manuscripts.[10] Despite that acknowledgment, Copernicus is given credit for originating the heliocentric theory. The elliptical nature of orbits was another early Greek idea which would have supported the earlier formulation of the heliocentric theory; Hipparchus verged upon that with his theory of "eccentrics" proposed in the second century B.C.[11] But new ideas that the earth moved around the sun, and in elliptical orbit, threatened the ancient Greeks' belief in the earth as center of the universe and the perfectionist need for circular orbits.

At the time of Greek speculation as well as in the time of Copernicus, truth could be sought and found to be acceptable as long as it did not conflict with religious or cultural beliefs on order in the universe. That left the way open for theories showing common sense, based on agreement with current knowledge, and posing no threat to current beliefs.

All Things Are Numbers

The ingenuity of the heliocentric theory eclipses a number of other noteworthy achievements of the very early Greeks, including calculation of the distance between the sun and the earth. The possibility of making such calculations stemmed from a belief that once the "number" in things was discovered, one could increase one's understanding and control over the world. Believing that mathematics offered knowledge of a kind superior and more reliable than that gathered by any other means, early

Greek philosophers sought answers to the basic questions of life in mathematical descriptions of the universe.

The concept that "all things are numbers" is credited to Pythagoras, who is thought to have been influenced by visits to Egypt and Babylonia as a young man.[12] The Pythagorean school he subsequently developed combined a love of beauty and symmetry, which formed the basis of contemporaneous philosophical and religious beliefs, with the idea that "all things are numbers." The Pythagoreans' mathematical calculations provided quantitative spatial arrangements of the universe, while their metaphysical commitment to beauty and symmetry kept the earth as the center of the universe. Later, Plato, influenced by the Pythagoreans and adding nothing new to their understanding of the universe, impressed on thinkers for centuries the importance of the idea that "all things are numbers."[13]

Although the early Greeks are generally considered the most important fathers of present-day mathematics, the value of their contributions has been controversial at one time or another and their influence viewed as a mixed blessing. For example, the Pythagoreans are criticized for developing the geometrical process but not applying it, leading to a separation of pure and applied science that delayed progress in science and technology for centuries.[14]

As another example, Plato believed that theoretical science, developed from purely philosophical examination, was the only reliable means to gain knowledge of the universe. He did not believe that visual observation or use of the other senses was reliable, because all we can see in everything is merely an imperfect copy of the real design, which we can know only through contemplation. Such a limitation has been claimed to be more inhibiting than inspiring for science. The fact that Plato was not an observer and hence not an experimenter has been proposed as a reason for the decline of Greek science.[15]

In contrast, Platonism (with its Pythagorean heritage) is credited with inspiring the mathematical science of nature that later flourished in the seventeenth century. The scientific revival of the Renaissance was basically a break-away from Aristotle and a return to Plato.[16] If the pure and applied are indeed separated, with pure science not needing the applied for progress, and if the early Greek mathematical science could drive the renewal of science during the seventeenth century scientific revolution, there seems to be little basis for the argument that Platonism contributed to the decline of science in early Greece. Indeed, ardent belief that "all things are numbers" is more or less the basis for theoretical physics today.

Elegant Astronomers

Individuals who came later in the Greek civilization continued to contribute to our knowledge of the universe. Hipparchus, for example, calculated distances between the sun and planets more accurately than anyone before; his measurement of the length of a year is remarkably close to ours.[17] Probably his most spectacular achievement was discovery and calculation of the precession of the equinoxes. Another Greek, Eratosthenes, contributed a calculation of the earth's circumference astonishingly close to present day values.[18] Thus some Greek scientists left the Platonic world of unreliable senses and used observations to draw conclusions about the universe.

Hipparchus was the first to catalogue the positions of the stars. His work provided a foundation for the work of Ptolemy, culminating in the *Almagest*, the highest achievement of Greek astronomy.[19] Ptolemy's explanation of planetary movement was brilliant; although some of his major concepts, such as that of an earth-centered universe, were later proved wrong, his work was the basis for all astronomy in the western world for the next fifteen

centuries. His technical inability to make precise measurements and his overdependence on the error-filled observations of others were the major sources of flaws in parts
of his work.

Ptolemy produced his "masterpiece of mathematical
artistry" in the second century A.D. It seems unrealistic to
believe that the observational or experimental aspect of
Greek science at an earlier age had been weak and
growing weaker. Ptolemy's work is evidence, rather, that
science reached a zenith at the end of the Greek golden
ages. Fortunately, Ptolemy's work was preserved to become an essential resource for the important successes of
later scientists such as Copernicus and Kepler.

The Father of Scientific Method

An early foundation for present-day biological sciences
was laid in the fourth century B.C. by Aristotle, considered
the most influential Greek philosopher and scientist. As
has been noted, Plato's material universe was only an
unreliable replica of the real design, which led him to try
to understand the real universe by theories arising from
contemplation. Aristotle, on the other hand, believed
physical matter to be real; its appearance to the senses
could be studied in order to understand the real world.
Critical of theorists satisfied with mental blueprints of the
universe, he proposed and practiced observation as necessary to know this world. Aristotle's scientific method arose
from what may be his greatest contribution, the establishment of laws for reasoning; those laws detailed propositions, fallacies, correct reasoning procedure, and a *deductive* system of formal argument (syllogism).[20]

Aristotle has been criticized for establishing a deductive
way of thinking that allegedly interfered with any growth
for science until the scientific revolution, when there was a
return to the methods of Plato.[21] Such an analysis seems

strange, since Aristotle argued for the scientific methods
of observation and experimentation. Even though he set
forth a system of deductive reasoning in his philosophical
treatises, Aristotle actually followed methods of *induction*
in studies where he collected empirical data from observations and experiments. According to Diogenes Allen:

> Objects which are sensible are the first to be known by us.
> Our knowledge starts from sense experience, that is with
> the particulars, and finds there the general (by means of a
> knowledge of essences). We reason inductively. This is the
> crucial point in all of Aristotle's philosophy . . . . But . . . the
> ideal is a deductive system of knowledge in which the
> particular is demonstrated to follow by necessity from the
> general.[22]

Despite his limitations, Aristotle is considered the father of
scientific method.[23]

Ageless Ideas

Aristotle's views of mathematics, astronomy, and physics
have in part proved to be wrong or unimportant, but for
centuries they offered plausible explanations consistent
with what was known and believed at the time. Interest-
ingly, he is criticized for delaying the progress of science;
his explanations are said to have prevented truth from
being known for twenty centuries. Aristotle's ideas were
accepted for so long because, at least in part, they offered
a reasonable and common-sense explanation of astronomy. Much later they were used to develop a theory
unifying human understanding of the physical universe
with the Christian religion. In the thirteenth century,
Thomas Aquinas changed theological thinking so that it
came to terms with Aristotelian teaching; the essence of
the change was that the invisible things of God could be

seen through his visible creation. That union is blamed for the perpetuation of Aristotelian dogma, because any attack on Aristotelianism was considered an attack on Christianity. Aristotelianism had persisted for some fifteen centuries without the support of Christianity, however.

A more important reason why Aristotle's "errors" survived so long is that no one came along to challenge them. Even for those concepts challenged, history shows that when they were important innovations in thought, they invariably survived the challenge of conflicting religious or cultural dogmatism. People were also not afraid of introducing new ideas because of a fear of challenging the establishment. For the early Greeks, the challenge provided by developers of the heliocentric explanation of the universe had threatened the established order and moral structure of the world long before Aristotle.

Aristotle's observations led him to propose some novel ideas about geological evolution. He described how the sea was replaced by land and land by the sea, resulting in the appearance and disappearance of a number of nations and civilizations throughout human history.[24] He explained such changes on the basis of natural causes operating in certain patterns rather than by supernatural forces. Only recently has natural science provided a more comprehensive understanding of geological changes, with much of the empirical evidence requiring very sophisticated technology.

As a scientist Aristotle's greatest contributions were in biology. *The History of Animals* was his supreme work and the most impressive scientific work of fourth-century B.C. Greece.[25] It was hardly improved on for over twenty centuries. Aristotle showed extraordinary breadth and reliability in his observations on animals. Some of his observations of animal behavior were not confirmed until the last century. He coined the basic terms *genus* and *species* used for biological classification, and so began the categorizing of living matter based on similarities and differences. Aristotle described the graduated differentiation of the animal and plant kingdoms known to present-day scientists, which can be seen as the basis of a rudimentary scheme of evolution.

For his achievements in studying various species of animals Aristotle could be called the father of zoology.[26] Yet his errors in biology are given as much attention as his countless accomplishments. For his time, Aristotle's achievements were extraordinary by anyone's standards. Again, the interesting question is why understanding in the biological sciences had to wait so long before anything substantial was added to Aristotle's work.

Aristotle did not appreciate a reduction of the universe's comprehensibility to numbers, nor value mathematics as much as some of his predecessors did. Yet his concepts of *continuity* and *infinity* recur in the subsequent mathematical works of Archimedes and Apollonius in the third century B.C.[27] The same concepts later helped scientists like Newton and Leibniz in devising infinitesimal calculus in the seventeenth century A.D.

Is Anything New? "Reinvention of the Wheel"

Twentieth-century marvels of human reason have included concepts of space and time typified in the ideas of Albert Einstein. Yet Bertrand Russell states that "on space and time, the theory of Aristotle has much in common with modern views."[28]

Aristotle evidently attempted to give time a dimension such as those of space, which would have been a unique concept.[29] His principal goal in doing so was to demonstrate the structure and properties of time as understandable in terms of the structure and properties of spatially extended magnitudes. Time cannot elapse unless there is change, which means that all elapse of time must be perceptible. Any change is dependent on magnitude since a change must be primarily a change of something having spatial magnitude. Thus, the parameters of time are dependent on change, which is dependent on the parameters of magnitude. For Aristotle time was merely a measurable quantity. It was continuous since it was a number belonging to something with a measurable magnitude that was continuous. Although Aristotle's so-called grand unifying design to describe a space-time continuum was unable to show all temporal entities and structures as derivative from spatial ones, the concept was there.

Matter has an inherent tendency to degenerate, a change that was equated with time by Aristotle. Such spontaneous degeneration illustrated an asymmetry in physical processes with regard to the direction of time, thereby anticipating the second law of thermodynamics and modern statistical mechanics.[30] For Aristotle the time changes were determined by a balance between metaphysical, form-giving, stability-maintaining forces from outside the world and an inherent tendency of matter to degenerate if deprived of the sustaining forces.

The essence of Aristotle's concept of time is that something in time, or measurable by time, must be changing. From that conclusion followed the statement that what is unchangeable must be necessarily everlasting and not in time. For Aristotle, God and the universe were among things that were unchanging. That made them everlasting things upon which time could not act. Thus Aristotle suggested that God, being unchanging and outside time, must abide outside the changeable part of the universe.

If time cannot elapse unless there is change or movement and if all elapses of time must be perceptible, under what conditions could measured time, the rate of its elapse, change? Time as measured could change if the matter observed or the observer recording time travels at a speed equal to or greater than the speed of light. The

movement representing change could not be observed or
recorded, because the light bearing the image of the
observed would never reach the observer. If something
moves faster than the speed of light it will not appear in
our universe if we require light to perceive it and recognize change or movement. Therefore such a rapidly
moving thing is timeless and, according to Aristotle, like
God and outside our universe. That raises the question of
whether something timeless can be outside our universe in
that sense yet still be within it in a different form,
imperceptible to our senses and measuring devices. Again,
as judged by Bertrand Russell, little has changed in our
concepts of time during the span of twenty-five centuries
from Aristotle to the present.

The understanding of time is thought to have made
progress toward truth by mathematical descriptions of two
phenomena by Einstein. One of them postulates that the
measurement of time is altered when the movement of a
time-recording device is different from the regular movement of the earth, in contrast to the time experienced by
clocks and observers stationary with respect to the earth.[31]
Observations do show that time-recording devices "run
slower" when traveling at high speeds relative to the
earth's motion. The second postulates that gravity changes
the measurement of time, so that time-recording devices
placed and observed at high altitudes "run faster" than
clocks at sea level, which are subject to a relatively greater
attraction of gravity.[32]

It is debatable whether the two phenomena represent
real "changes" in time. Experiments conducted to verify
the truth of these mathematical predictions may be unreliable because, as it is now understood, a recording device
or the mere act of observation can change the parameter
being measured.[33] In addition, changed conditions of
motion and gravity can change function in a measuring
device, altering the measurements it records. In general,

people are reluctant to accept any theory or explanation of natural phenomena that defies common sense. That has been true for the modern theories trying to examine old questions of time in a new way.

Other Aristotelian concepts also provided an early basis for modern scientific ideas. For example, Aristotle stated that it is impossible for one to attribute motion to a body by observing it at any one instant.[34] That principle has something in common with the twentieth-century concept of Heisenberg that one can specify the position of a body, or its speed and direction, but it is impossible to do both. Was that idea unique and original to Heisenberg? That is uncertain because it is known that Heisenberg was well schooled in and had a love of the Greek classics.[35]

Philosophy: The Unprovables, Are New Ideas Better than the Old?

Aristotle is criticized for denying the concept of *void* in the universe. Conceptualization of "void" or emptiness has always been a perplexing problem. Current arguments for the existence of void are based on the modern physical description of submolecular structure, showing the mass of all particles in a molecule (or in the universe) to be a small fraction of the total size of the respective realms, with the remainder of the space constituting a void.[36] Relying on this description and an intuitive concept of void, it is not difficult to accept the criticism of Aristotle's belief that there is no void.

Aristotle claimed that the nonexistent void would be a kind of place, a non-bodily, three-dimensional, spatial extension independent of any bodies that might happen to occupy it and which would be receptive of occupying bodies; that is, it would remain in position even when occupied.[37] For Aristotle it evidently was not possible for particles to be independent of any void-like space they occupied. Likewise, today it is difficult to see how the space occupied by the structural particles of an atom could be independent of their presence and sphere of movement. The space they move in may be likened to a void but that void is essential for their pathways of movement, without which there would be no atomic structure. But Aristotle also stated that motion is impossible in a void; if that is true the space through which particles move is not a void. Motion is impossible because "the void is too negative to be the carrier of distinctions of directions."[38] Fields of force are determinants of motion and direction for particles in a space. Such a space is not a void with the absence of particles, because with the fields of force motion is possible.

The concept of no void can be represented by all space being occupied by particles of material having a defined space as their sphere of activity and necessary motion; with that space providing pathways for essential movements of the particles on a regular basis, no part of that unit of space can be considered a void or to contain a void. To say that pathways exist for particle motion through a space denies it as a void. True void would be a space with no particles ever present.

The pathways for particle movement describe a *field*; for some scientists such fields represent a truer reality than the particles that may occasionally pass through.[39] With fields gaining reality there can be, as Aristotle believed, no void in the universe. Thus the criticism of Aristotle for his disbelief in a void may be unjustified. There is merit in his concept, although his arguments for it may have been flawed.

A Pioneer of Theoretical Physics

Aristotle also deserves credit for his largely overlooked concepts of dynamics, which make him a pioneer of theoretical physics.[40] He envisaged empirical sciences (including optics, harmonics, and mechanics) as being heavily dependent on and even subordinate to mathematics. Although his development of dynamics was hampered by lack of differential calculus, Aristotle was able to (a) provide for acceleration; (b) construct what is in effect a law of inertia; (c) describe a "motive power" which may be equivalent to the concept of momentum; and (d) anticipate a version of Isaac Newton's Third Law that action and reaction are equal and opposite.

Science: Why a Great Beginning?

If its foundations were provided by the early Greeks, why did science not flourish during the next twenty centuries? It may be that people did not need science and its applications at that time.[41] That answer is based to some extent on the fact that industry had slave labor to perform most of society's tasks. Only with the abolition of slavery would an interest in technology, to replace that labor, begin. It would follow that when people began a conscious effort to reject slavery, an interest in developing technology would become essential in order to yield the fruits of labor, agriculture, and industry essential for human existence.

The early Greeks and other ancients had few aspirations for practical applications of scientific knowledge; they pursued science rather for the joy of discovery. In fact, the development of science primarily as a basis for improving technology is something not consciously pursued until the present century.

Yet the early Greeks must have understood very well that practical technology could be a product of science. Archimedes demonstrated the connection by using his knowledge of specific gravity to answer a question on how much gold an object contained, and by applying his knowledge of mechanics to help his city withstand the onslaught of an invading army. It has been suggested that, "but for the abundance and cheapness of slaves, Archimedes might have been the head of a veritable industrial revolution."[41] With the later disappearance of slavery, however, scientists continued to work for the joy of discovery long before they worked for any technological benefits.

There are no adequate sociological explanations for the appearance of science, pursued with such unusual brilliance and merely for the joy of discovery, at a relatively early time in the history of human intellectual achievements. Can the lack of significant progress for twenty centuries following the early Greeks be adequately explained sociologically?

Chapter 3
A Renewal of Interest in Science

Cycles of Interest in Science

The sixteenth and seventeenth centuries marked the beginning of an era of activity called "the Scientific Revolution."[1] If we could understand the reasons for that important renewal of scientific achievement, we might have more confidence that human effort could be directed to eventually reveal the secrets of the universe—if the causes of the Scientific Revolution can be attributed to human efforts. Even if we cannot fully explain the causes of the Scientific Revolution we may gain some insight into why science did not advance for so many years until that revolution began.

The remarkable innovations of the Scientific Revolution were not necessarily achieved by making new observations. The innovations were to a great degree simply new ways of looking at old problems.[2] In general the questions had been asked for many centuries before the Scientific Revolution. Consider Copernicus' resurrection of Aristarchus' heliocentric theory; Copernicus' contribution was not based on new observations or experiences. Even at a later time in the nineteenth and twentieth centuries the great controversies relate to problems already known to Aristotle.

Modern scientists sometimes introduce concepts that appear to be completely new and revolutionary but are in reality merely reexaminations of old questions. Are any truly novel ideas generated today, or are all "new" concepts merely restatements of the old? In Plato's view, an idea or concept never changes; it is not born nor does it decay. It comes into peoples's minds at times and at other times is not thought about.[3] Lacking awareness that new concepts may be no more than sophisticated restatements of the old, we can easily fail to understand how for centuries people seemed not to comprehend "what should have appeared so obvious to them."

Perhaps a drive to reexamine old problems is merely an episodic feature of history, with science flourishing in eras such as the time of ancient Greece and lying dormant in times such as the Middle Ages. Such a simplistic view of historical cycles provides no explanation for the inspiration behind the Scientific Revolution. Pagan and Greek concepts of time tended to be cyclical. In contrast, the Old Testament portrayed human history as linear because from the beginning God was directing it along such a path.[4] For the individual scientist, interest in science with its joy of discovery is more linear than cyclic. During the centuries of stagnation in the Middle Ages, lack of achievement can hardly be attributed to a want of scientific interest and motivation.

Did Religious Doctrine Stifle Scientific Progress?

Was the desire to reexamine old problems stifled by religious interests that looked on scientific discovery as a threat to established religious doctrine? Most scientists during and immediately following the Scientific Revolution were deeply committed to religious faith; their faith did not interfere with their scientific achievements.[5] Pressure from religious doctrines perceiving scientific concepts as a threat had less deterring effect on the advancement of science than did academic conservatism and hostility to new ideas.[6]

Contemporary scientists are probably not much different from others throughout history. When William Bayliss and Ernest Starling reported on secretin, the first hormone to be discovered, they were severely criticized by fellow scientists for suggesting a drastically different mechanism for the regulation of functions in the mammalian body.[7] Edwin Burtt said of Copernicus' heliocentric theory:

> Contemporary empiricists, had they lived in the sixteenth century, would have been the first to scoff out of the court the new philosophy of the universe.[8]

Despite such critical responses to novelty throughout history, scientists do at times accept, with few reservations, new theories built from minimal shreds of empirical evidence. Their acceptance is frequently based on unscientific criteria such as the beauty of an idea. (For some, beauty represents truth.) Thus, some factors that may be inherent to human nature mount resistance to the progress of science while others tempt people into accepting unreasonable theories as approaches to truth. Other factors, including what might even be called superstition, retarded the development of science more than did any opposition by the Church.[9]

Historically, the zenith of conflict between science and religion was between Galileo and the Roman Catholic Church. That conflict over removing the earth from the center of the universe was not inevitable; with a different leadership the Church might have found Galileo to be right.[10] Not until this decade was that dispute officially resolved. The fact that the Catholic Church maintained its position for so may centuries, despite an overwhelming scientific consensus, leads many to conclude that real conflict continues to exist between science and religion. (Some religious leaders are inclined to see scientific thought as an adversary, regrettably.)

Scientists are driven by inherent factors generally stronger than external cultural influences, so their work is seldom influenced by any religious establishment. They may hold an unwavering faith in what their investigations have convinced them is true; often that faith is held despite inabilities either to make accurate enough observations for mathematical proof or to accumulate unambiguous supporting facts. Confidence in scientific method and theory can be stronger than faith in formal religion or other factors that influence human thinking. Science can motivate people for a lifetime, even with little promise of reward.

Rather than religion interfering with scientific progress, the major influence has been in the opposite direction. The achievements of science have given people apparent reasons to abandon religion as a source of knowledge and understanding. In our time, the change of allegiance from religious belief to a belief that science can make one self-sufficient has proceeded very far. For many individuals, a mathematical science and modern biology replaced God as the most promising source for explanations of human existence and destiny.

The renewed interest enabled a science of numbers to become again the "sole authority" in a new kind of religion because the mathematics of proportion provided a universal key to the design of the universe.[11] Today that design is considered to be hidden in a kind of cosmic code.[12] In addition, "the principle of proportions was also found in the structure of the human body and in the newly adjusted function of men's moral existence."[11] Mathematical science seemed to give people new powers over their environment and to put the fruit of the tree of knowledge once more within reach.

Mathematicians with a lineage back through Plato to the Pythagorean school may argue that mathematics did not bloom until liberated from the bondage put on by the amalgamation of Aristotle and Christianity by Thomas Aquinas. The opportunity for freedom to flourish was always there, however; further, Aquinas did not appear and develop his constraining theology for more than fifteen centuries after Aristotle. Liberation from religious restraints was not necessary for a return to Plato's mathematics.

Christians seldom recognize Aristotle as an important person used by God for achieving goals in science. Aristotle is considered a pagan; moreover, when his works were rediscovered by Arabic scholars, they were used to fashion Islamic logic and science. Christians must accept God's use of such individuals (and their influence to produce a conflicting religion) for bringing progress to humanity. Further, Aristotle was the antithesis of Plato, whose works influenced the Christian theology developed by Saint Augustine. Despite that, Thomas Aquinas, an important thirteenth-century Christian theologian, gave Aristotle the seat of authority on natural philosophy. Several centuries later, Aristotle's loss of influence was hardly a sufficient reason for Christian thinkers to initiate a scientific revolution.

To the question of why science slumbered for over twenty centuries, an answer blaming religion is an answer unsupported by the facts.

Science Develops Despite Mysticism

Throughout much of history, many people have believed their life decisions and actions to be determined or affected by a variety of mystic forces, encompassing the "seen" in stars and planets and the "unseen" in angels and demons.[13] Such an entrapment in metaphysical beliefs may seem absurd today, but the fact is that many otherwise modern individuals continue in such beliefs. Current "sophisticated" cultures support all kinds of ancient superstitions; an obvious example is the regular appearance of astrological predictions in daily newspapers. People continue to believe in horoscopes, the occult, and many other forms of mysticism in their attempt to understand the unknowable. Even early scientists following certain metaphysical practices were able to make contributions to the Scientific Revolution. Following a rather rigidly defined scientific method they were able to keep the metaphysical from influencing their scientific endeavors.

Astrology, a centuries-old practice in mysticism, claims to enable people to foretell the future by studying the supposed influence of the relative positions of the moon, planets, sun, and stars on human affairs. Such an influence would make everything in one's life predetermined, because all the positions of these heavenly bodies are predetermined.

Casting off astrology is an important step in freeing humanity from determinism so that one's fate is determined not by the heavens but by a person's own will and actions. Human destiny decided by human beings gains dignity. Life is not merely a pawn in some chess match of the stars.

The determinism of metaphysical or superstitious beliefs could influence the advancement of science. But in Protestant cultures, where predestination was important, belief in a completely determined life seemed to have no adverse effect on science. Rejection of a metaphysical basis for a determined existence also seems to have had little effect on people's motivation and action to promote science. Science continues to offer the promise of a complete understanding and mastery of human existence; yet many people who accept science as a tool for mastering their world continue to seek metaphysical or religious answers to their deepest questions.

A Remnant of the Essence Is Conserved

Motivation to pursue science in the Western world did indeed exist before the beginnings of the Scientific Revolution. Arabic scientists were as driven as those in earlier cultures and made some of their own contributions. Although Islamic science made no astonishing leaps forward, the Islamic world is recognized for its important contribution as the conservator of Greek scientific thought. Indeed, that was the major channel through which Greek thought was saved and passed on to the Christian world.[14]

Explanations for lack of progress by Arab scholars generally follow the same arguments for the lack of progress during the domination of society by Christianity. One suggestion is that the monotheism of Islam, Christianity, and Judaism made it seem more like heresy to think about natural causes for explaining the world.[15] But in addition to interpreting, analyzing, and commenting on what they inherited from Greek science, the Arabs did make a number of original contributions in astronomy, mathematics, physics, and biological sciences.

In addition to the Arabs the Romans also contributed by saving a body of Greek scientific knowledge; what they saved became an important source of information for scholars of the Renaissance.[16] Although the Romans had the foresight to act as conservators, they showed little interest in seeking further knowledge through science. During the time of the Roman Empire they were a practical, technologically-oriented people who were not

given to intellectual speculation nor ready to support
scientific experimentation.

Experimental Sciences' Beginnings

The achievements of Copernicus or Galileo are commonly believed to mark the beginning of the Scientific
Revolution. Scientific advances appearing earlier in the
Middle Ages are generally ignored because they were
usually built on the rediscovered ideas of Aristotle. The
gate to the Scientific Revolution was not yet opened by
those who were subsequently to discard Aristotle's thinking in favor of the Platonic tradition.

Einstein stated that the development of Western science
was based on two achievements of mankind, (a) the early
Greeks' invention of a formal system of logic in Euclidean
geometry and (b) the possibility of finding causal relationships by systematic experimentation, beginning at the
Renaissance.[17] His assertion is consistent with the hypothesis that Greek science declined because such scientists as
Aristotle and Plato were not experimenters; any weak
inclination to experiment grew even weaker with the
passing of such leaders.[18]

The idea that methods for experimentation were not
conceived before the Renaissance is unfounded, however.
The Middle Ages produced at least one important scientist
who, under the influence of Aristotle, not Plato, wrote on
the nature of scientific inquiry long before the beginning
of the Renaissance. During the twelfth and thirteenth
centuries Robert Grosseteste stated that science begins
with people's experience of phenomena, followed by an
endeavor to discover the reasons for the experiences.[19]
The causes are sought and their agents are analyzed.
Using these assessments the observed phenomena are
reconstructed in the form of a hypothesis. The hypothesis
is then tested and verified or disproven. Grosseteste thus

established a basis for experimental science. His Aristotelian-based scientific method was developed centuries before the Renaissance and before the beginning of the Scientific Revolution.

Opportunities Begun with the Renaissance

The Scientific Revolution may have been set in motion by the release of humanity from restricting influences other than formal religion. Sweeping changes in people's general outlook on life and the world stimulated the fifteenth-century Renaissance.[20] The influence of religion persisted, but human interests came to predominate over divine interests. "Humanists" popularized the entire body of ancient Greek learning and literature, introducing scientists, philosophers, and artists to a whole new world that surpassed their own civilization. Awakened thinkers discovered objective and universal laws in nature, history, and human nature, but especially in mathematics. Hope arose that universal laws would replace arbitrary, inexact, subjective, and often oppressive laws based on intuition and narrow insights.

The cultural flourishing at the time of the Renaissance is considered remarkable but its recorded achievements document little that was fundamentally new.[21] The people of the Renaissance adapted the skills of the East to the conditions of Western Europe, giving an appearance of dramatic change and technological advancement; the Europeans' sense of achievement was reinforced by the period's economic and political developments. Newly found skills changed methods in manufacturing, mining, building, and farming, thereby slowly revolutionizing the Western way of life. It could be argued that the new skills had no great influence on European thought or aesthetic perceptions. The essence of the Renaissance, as a release from the past leading to development of a Scientific

Revolution, may in reality resemble the change in some Third-World entities today, catching up with the Western world and learning to employ its achievements.

Did Technology Drive the Scientific Revolution?

Western civilization continues to consider many things unique about its people to account for the Scientific Revolution. Western Europeans are described as accomplishing the revolution by inventing new habits of mind and new methods of inquiry, ingredients which blossomed fortuitously in the seventeenth century and which neither the ancient world nor Byzantium could have achieved. These new elements in the equation ended domination by civilizations of the Mediterranean basin and produced a dissolution of old forms of society, a decline in the role of the Greek and Roman heritage, and a waning of the influence of Christianity.[22] Was the genius springing from Western Europeans unique, and of a magnitude and importance never experienced before? One reputed unique-to-the-Western-world prerequisite for science to flourish was the development and refinement of technology, based on the craftmanship of European artisans in the Middle Ages.

Indeed, before science could flourish, technical skills had to reach a certain sophistication to produce such items as instruments for making precise scientific measurements. The fact is that such technical resources were available to Eastern civilizations long before their application in Western society, although their presence in the older cultures was not associated with noteworthy scientific achievement.[23] Superior technical skills of artists, artisans, and builders did contribute to the success of scientists in the Western but not the Eastern civilizations. Skilled technicians from the East produced textiles, mosaics, ceramics, glass, and metal work of remarkable quality.

Products appearing even before the time of Christ were of a quality far surpassing anything from the West. In addition, the ancient East is credited with producing many drugs, dyes, and other chemicals; precision instruments; building techniques; advanced agricultural, mining, and transportation practices; innovative arms and armor; gunpowder; paper; printing; canal locks; rudders; and the mariner's compass.

China in particular is responsible for three developments or discoveries that have changed people's lives as much as anything else in history.[24] The most significant was printing, using nonmovable type in the ninth century and movable type in the eleventh century. That was four hundred years before the appearance of printing in Europe, where printing has been considered a discovery of Western civilization, and a development necessary for the beginning of the Scientific Revolution.

China's second great achievement was the invention of gunpowder; the third was development of the magnetic compass, before which exploration of the globe was virtually impossible. The Chinese were also adept in developing time-keeping devices and seismographs, and in casting metals and making porcelain.

If technology was critical to initiating a scientific revolution, what kept a technologically more advanced Eastern civilization like China from developing scientifically? Some believe that China's failure to develop Euclidean geometry was responsible.[25] Euclidean geometry is one example of universal laws, but the Chinese, not conceiving of a divine law-giver, did not seek universal laws, the basis of a scientific understanding. Others suggest that the centralized bureaucratic structure of Chinese society did not allow people to break out of the mold of existing orthodoxies and develop the independence of thought critical for science. It seems doubtful that Eastern orthodoxies were any more restricting than the Western world's religion, customs, superstitions, and governments so often claimed as impediments before the Scientific Revolution. The Chinese constraints may appear to be more restrictive only to a Western mind.

Compared to scientists of medieval Christendom, Chinese scientists conducted studies that were far more comprehensive, and their ideas were more likely to be followed through to completion.[26] Another contrast concerns one of the outcomes of the Western Scientific Revolution; a reductionism, examining the universe at the level of its basic subatomic building blocks. This philosophy, associated with a Western approach to nature, could be considered more destructive than the Chinese holistic attitude to the universe.

China's most important technologic accomplishment, printing, could actually have been developed elsewhere even earlier. The essential elements for that achievement were available in the Western world from antiquity.[27] The technologic essentials included punches, molds, lead castings, presses, paper or its equivalent, and ink or paint.

Printing may have been as revolutionary to human thinking as computers are today for organizing and storing information far in excess of what the mind is capable of processing, storing, and recalling. A profound reorientation of the processes of thinking may have been necessary to realize the value of producing and storing thoughts on paper.[28] The technology to produce computers was similarly available long before all the presently known possibilities for computer usage were realized.

The appearance of universities, the development of a form of logic that was more highly quantified in the scholastic tradition than in Aristotle's time, and the organization of thought according to spatial models prepared the way for the Gutenberg press.[28] With the appearance of printing, lectures were no longer the only means available to university students for acquiring new information.

Throughout history technological development has been the work of skilled artisans; only in this century have science and technology become closely related.[29] Today, technology presents problems that stimulate scientific investigation. Also, scientists develop a need for new instruments which are then produced by technicians, but this is a recent relationship that did not always exist. For example, did Galileo's observations of the heavens to verify the heliocentric cosmology of Copernicus have to wait for the technological development of the telescope? The fact that lenses could be combined as in telescopes and microscopes to magnify the image of a distant or small object was known to Roger Bacon in the thirteenth century, more than three centuries before Galileo's use of the telescope.[30] Copernicus also could have used that technology but instead he developed his ideas using observations from antiquity. The technology was there long before these scientists developed a need for it; a heliocentric theory could have been promoted earlier. Though in possession of perhaps greater brilliance and the same information some fifteen centuries earlier, Ptolemy did not hit on a heliocentric theory.

Throughout history few scientific achievements have sprung directly from technology. Until recently in Western civilization technology was more sophisticated than science, with little direct relationship between the two.

Is Science Pursued for Technology?

Arguments that science needed technology for its discoveries are matched by arguments that technology needed science for its progress. Until recently it could be said that "scientific discoveries do occasionally lead to applications in the form of new technology; this is rare . . ."[31] Even today, however, technology, which enables people to achieve some purpose, is not automatically created by science; science generally has not been a catalyst for the development of new technology. Of course the creation of new technology requires its creators to have scientific training and a comprehension of the nature of things, but technology progresses under its own momentum.

The Essence of Scientific Method

The Scientific Revolution, as a manifestation of rethinking human existence, included ideas about the significance of different scientific methods. Of all the possibilities, the primary requirement for science to appear and flourish seems to be an ability to make generalized conceptions of scientific explanation and to use mathematics for their proof.[32] This prerequisite was discovered by the ancient Greeks, who invented natural science. They assigned to the universe a permanent, uniform, universal, and abstract order governed by laws which can be understood and explained by deductive thinking.

The Greek's generalized use of scientific theory was based on principles of noncontradiction and the empirical test. This method of scientific explanation is attributed by many primarily to Euclid, whose *Elements* contains the form of logic used in his cataloguing of Greek geometry. It is understandable that the Scientific Revolution has been thought to be most importantly a return to such a mathematical basis of Plato and the Pythagorean school rather than an empirical basis usually attributed to Aristotle.

Euclid's logical ordering of all knowledge in Greek geometry up to that time was a forerunner of the current scientific method of progressive exposition and demonstration.[33] This method begins with propositions about what can be expected to be found by analytical methods, leading to diagramatic illustrations which prove or disprove the proposition. Conclusions follow which include new questions that might arise and problems that might be solved. These in turn lead to new propositions, suggesting new forms of explanation. The Euclidean mathematical propositions cannot be falsified by observation or any other empirical means; changes are made only by re-thinking.

The Greek scientific tradition also produced inductive methods of collecting and ordering data, of using data to establish causes for phenomena, thus providing a process for discovery. Early Greek methods for looking at the universe led to the use of hypotheses, to the development of criteria for accepting one theory over another, and to decisions on the place of theoretical concepts in scientific explanation.

The scientific methods of abstract conceptualization (as in mathematics), experimentation, and observation without intervention have been categorized into inductive or deductive forms by various philosophers. Some have generalized that the Scientific Revolution was not possible until deductive constraints were removed from the way people thought. But the hypothetico-deductive method, which is in essence the modern scientific method of discovery, is implicit in the Aristotelian method and that of Descarte and his followers.[34] A return to the Platonic school and its inductive way of thinking was deemed important to some.[35] For certain areas of science, however, Plato's concepts are viewed as a disaster; for example, only when biology was emancipated from Platonic thinking was it able to emerge as a science.[36]

If the scientific method cannot accurately be described as one of induction or deduction, the ancient Greek scientists cannot individually be credited or blamed for their contribution of one form or another, or for any overall success or failure of science.

History does not show any unification of the mathematical and empirical methods of the Greeks until the thirteenth century. At that time Robert Grosseteste and his pupil, Roger Bacon, using the theoretical empiricism of the twelfth century and the deductive form of scientific explanation learned from Euclid and from Aristotle's logic, proposed a concept of science that brought together the experimental, mathematical, and deductive approaches.[37] Overall, their scientific method was inductive; explanation was based on actual observations or experimental data. One proceeded from effects to a cause, the dependence of which was based on sense experience.

Grosseteste also understood that experimental methods could be used to deny or falsify hypotheses, but could confirm them only to some degree of probability.[38] Grosseteste thus anticipated two concepts generally regarded as twentieth-century innovations: (a) substitution of falsifiability for verifiability as the reliable means for evaluation of scientific explanation, and (b) development of statistics to assign probability to scientific conclusions for their acceptance as consistent and universal. The essence of these accomplishments of the thirteenth and fourteenth centuries was fundamentally similar to concepts of scientific methods developed during the sixteenth century.

By the sixteenth century, scientific thinking included analyses of the logical relationship between theories and observations, and of the criteria for falsifying or verifying a hypothesis. Also included was the Neoplatonic concept that all of nature behaved according to universal mathematical laws which could be discovered and analyzed by new mathematical techniques.

Were Philosophers Essential for the Scientific Revolution?

Philosophers might argue that a process of philosophical reorientation, such as described here for late medieval times, was as necessary to the scientific accomplishments of the seventeenth century as it was to that of the Greeks. But scientists seem indifferent to the attempts of philosophers to analyze and categorize various steps in the scientific process.[39] It may be that it is essentially impossible to define the scientific process in a precise way.

As a critical historical analyst, a philosopher can use the approach of one scientist or of a well-defined scientific school to produce a critical description of a particular scientific process. Different scientific disciplines use different methods, however, so philosophical generalizations are often suspect. Philosopher-scientists were able to document the scientific process at any one time, but is is more difficult to pinpoint those who set the stage by being innovative centuries before their time. It may be that development of the scientific process has been evolutionary rather than revolutionary, with no particular philosophical insights as contributory prerequisites.

Scientific philosophers do seem to believe that almalgamation of the Euclid-mathematical-Platonic concepts of nature with the empirical approach of Aristotle allowed Galileo and his contemporaries to reexamine the scientific achievements of Greeks such as Archimedes, thus beginning the Scientific Revolution.[40] The implication is that if the Greeks had had the benefit of scientific method developed 12 to 15 centuries later and used by seventeenth century scientists, they would have accomplished more. On the other hand, ordinary scientists may doubt that a particular system of thought had to be developed before success could be realized. Scientists do not learn how to do their science from philosophers and did not wait for their methods to be articulated by philosophers before making discoveries. It is true that the system put philosophers in a position to evaluate the achievements of science, with certain technical successes regarded as firmly establishing the general lines of subsequent scientific thinking.

At the beginning of modern science, both scientists and

philosophers were intensely interested in the process of scientific thinking, including such early scientist-philosophers as Grosseteste, Roger Bacon, Galileo, Francis Bacon, and Descartes—and of course Isaac Newton, who came later. By the seventeenth century, scientific thinking was enhanced by a realization of full interaction between theoretical science and scientific methodology, experimentation, and technology. The results of technical maneuvers contributed to the confidence of scientists in the more philosophical, theoretical science. At each stage of development philosophers credited themselves with improving the pursuit of science.

Francis Bacon, for example, developed a new version of induction to replace the method of syllogism evidently then perceived as bankrupt. He conceived a scientific method in which observation, classification, and experimentation would lead to theories which would in turn lead to more penetrating experiments and still deeper theories, until knowledge would be extended, perhaps even to the achievement of immortality.[41] Bacon's method was thought to make possible a total reconstruction of sciences, arts, and all human knowledge, which would extend the dominion of the human race over the universe. That method was neither unique nor novel, however, and it has not achieved the stated goal. The notion of induction in his version was used by Aristotle; in fact Bacon's version offered no advancement beyond Aristotle's. Regrettably, none of the philosopher-scientists contributed dramatically to elucidating and improving the process of scientific thinking.

A Rethinking of Old Ideas by Galileo

Galileo, born in the sixteenth century, created an important crossroad in the history of science; his achievements are considered a major factor in initiating the

Scientific Revolution. The accomplishments of Galileo were fabricated from important pieces of work from earlier scientists.[42] The principles of his theory of dynamics were originated by Buridan in the fourteenth century; Galileo learned of them through the writings of one of Buridan's pupils, Albert of Saxony. Galileo gave Albert's theory of dynamics a more precise mathematical description. Galileo's new theory also reflected the ideas of the medieval Muslim philosopher Ibn-Badja, ideas Galileo used to transform Buridan's theory into a generalized one. Galileo's great achievement was to fashion ideas known for several centuries into the fundamental principles of a new science that became classical mechanics. His medieval predecesssors did not attempt such a synthesis. Given the necessary background, Galileo is a supreme example of a scientist with the insight and imagination to synthesize a new theory by rethinking old ideas.

Today's Beliefs Become Tomorrow's Myths

One might think that if people in the late medieval time and more importantly in the seventeenth century laid the foundations for the Scientific Revolution, then the philosophic method and empirical evaluation employed at that time must have been "highly scientific," producing reliable and significant advances in science. Most of those accomplishments, however, especially in physics, are currently considered out-of-date beliefs; to some they have become myths. Each age seems to produce ideas which become myths for the next generation. When people use the same philosophical approach and the same kinds of methods today as were used to produce the myths of yesterday, present-day accomplishments also become myths, if not now, at least in the next generation.

If, on the other hand, out-of-date theories are discarded in favor of a new theory with a better explanation, and if that change of thinking does not make the old way unscientific, then scientific progress is not merely a measure of how much acceptable scientific information is accumulated but is a rethinking that provides better explanations than the old.[43]

The Pause in Scientific Progress

Many of the reasons given to account for the Scientific Revolution are unsupported by much evidence. There is little disagreement that it was a "revival," a time of rethinking of ancient questions, rather than a movement driven by introduction of brand new ideas or by an abundance of new scientific observations. The renewal of scientific interest was a revolution in the sense that it was a rather dramatic return to the philosophical and scientific ideas of the Greeks, but it was not characterized by acquisition of new habits or new forms of inquiry.

Two questions remain unanswered. One asks why there was an interval of many centuries before the Greek scientific basis was revived to begin a scientific revolution. For the other, if the early Greeks had the freedom of thinking that would later spark a renaissance, and if they had the scientific knowledge to provide the basis for a later scientific revolution, why didn't these two ingredients result in a continuing progression of science in early Greece to compare with what came many centuries later in the Scientific Revolution?

Reasons for the Pause

In general, most accounts of the Scientific Revolution credit the seventeenth-century scientists with far greater originality and achievement than were possessed by thinkers of the early Greek culture. Also, for a revolution to occur people had to make progress in many areas of their existence, such as craftmanship, government, economy, and the development of reason; they also had to escape from any restrictions of superstitions and religion. Such accomplishments are commonly regarded as products of human efforts in growing to realize our ultimate human potential.

The success of early Greek science, followed by a long period of decadence and subsequently a renewal in the Scientific Revolution, are regarded by humanists as immanent to human history.[44] A cycle of success, inertia, and success, and the timing of such a cycle is not seen by humanists as representing introductions of changes from without according to some divine plan. The events of such a cycle are believed to symbolize humanity's development, virtue, decadence, and revival. Such cycles are considered inherent to human development; to the humanist they need not be rationalized, only described.

For humanists, scientific progress and indeed human progress are inherent, evolutionary, and without a single identifiable cause; each step in their development maintains a continuity from the simple to the more complex. Each evolutionary step depends on an apparently random, indeterminate, and possibly arbitrary element, compounded of unique personal factors and historical accidents, or perhaps what people call "luck" when it seems in their favor. Although most humanists abhor metaphysics, such a description is really a rationalistic, metaphysical "explanation."

The cycle of human history from the early Greeks to the Scientific Revolution hardly seems evolutionary, however. In an evolutionary process, decadence cannot be explained as a part of any scheme, whatever is said about ultimate purpose. Evolutionary processes have an inherent "time's arrow," with continuous feedback providing for a progressive development based on testing and selection of numerous variations. Such progressive evolutionary development was not evident during the period of time from Greek antiquity through the Scientific Revolution.

In summary, no valid rational explanation has been advanced for the cycles in scientific achievement. Our own suggestions concerning the decline of scientific achievement in early Greece and the later rebirth of science will be offered later.

We now turn to a consideration of how people approach and answer questions; how was the rethinking done that produced the Scientific Revolution?

Chapter 4
Rethinking

Functions and Expectations of Scientific Method

Scientific method, although difficult to describe, has some important and unique functions.[1] One is to validate general statements as true or false, in probability terms based on statistical evaluation. Another function is to describe matter with regard to its simplest basic components, as is done in reducing biological functions to the language of chemistry and physics. A third function is to establish the cause of each event, as when chemical information allows one to explain the basis or "cause" of a biological function. Western scientific method is based on this *reductionist* approach, reducing or dissecting everything down to its smallest subunits with the idea that a full understanding of causes will ensue.

Scientific Method; Are Descriptions Illusive?

Definitions and analyses of scientific method have been illusive. In the past much confusion stemmed from labeling scientific methods as inductive, deductive, or something else that identified an individual method with an early Greek. Such generalizations are seldom of much value, especially since confusion is so evident about the proper relationship of laws, theories, observations, data, and conclusions in natural science.[2]

Scientists believe the most important criterion for scientific method is an objectivity in which an experience or generalization is accepted as valid by all. Scientists cannot view scientific aspects of the universe with complete objectivity, however; they never completely abandon the subjective, unexplainable aspects of individuals and cultures. Every experience of one's life is recorded in the conscious or unconscious and subsequently can influence every decision made and idea formulated. That includes scientific theories which require objectivity. Furthermore, for experience to be possible, indispensable conceptual presuppositions or preconditions are necessary and they can differ as each individual differs. Each scientist therefore describes scientific method differently, for each scientist's method is based at least partially on the unconscious and subjective.

The Inductive Approach to Science

Induction is generally given credit for the successes of science since the seventeenth century. In inductive thinking, information is gathered on many particulars and, from those particulars, general statements or conclusions are made. Induction is not a method of logic that can lead with certainty to *truth*; it leads only to probabilities.[3]

Accumulation of information on particulars does not assure any greater understanding. In fact, the inductive scientific method, followed to seek understanding, most often leads to error or to nothing.[4] Most scientific endeavors achieve little in the way of understanding, partly because inductive theory insists on the primacy of facts (simple and uncomplicated recordings of the senses), counting on the results of experience as an essentially reliable record of the human senses.

Why does the inductive approach so frequently fail? Because "everything that reaches consciousness is utterly and completely adjusted, simplified, schematized and interpreted"[5]; "innocent, unbiased information is a myth"[6]; and "experience is itself a species of knowledge which involves understanding."[7] Thus, the facts are not simple and uncomplicated recordings of the senses; the "facts" are not the same for all people. Any "primacy of facts" applies only when the facts, as "products" gathered by the senses, are critically scrutinized and recognized as being influenced by all the experiences of the observer. Inductive methodology does not aid in deciding which observations are preferable to others, making their selection dependent on a decision about whether the collected facts are leading somewhere or nowhere.

As with any system that cannot lead to truth, inductive thinking will be justifiably preoccupied with real or imagined problems of discovery and verification—that is, with criteria for judging an idea right or wrong. That becomes academic for most theories, which are modified, are incorporated into new ones, or merely fade away. Some theories disappear without any chance for modification because they are seen to be in error from the beginning. Inductive theory provides no acceptable explanation for the origin and prevalence of error generated by this method of scientific discovery.[8]

Even if scientific methods could be described as inductive, deductive, or a combination of the two, scientists do not learn any one such method before research can begin. In fact, they usually have little interest in being able to articulate the steps they take to realize the success of discovery. Scientists are described by some as working close to the frontier between bewilderment and understanding.[9] How can something so imperfectly understood be explained and classified?

For the most part, philosophers have not been able to

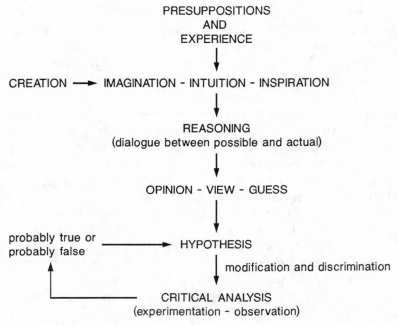

Figure 1. Schematic of Scientific Process.

agree on descriptions of how scientists function or on how the *should* function.

Science Begins with Experience

The scientific process begins with experience, as shown in Figure 1. That experience most importantly includes all of one's formal education and previous scientific endeavors, but it also includes the sum total of one's mental existence. The mental totality supports the ability for greater experience and generates the many presuppositions directing choices in that ability to experience. Presuppositions are critical to human experience. Such experience is individualized since presuppositions are not immanent, not arising from one's experience here and now, but are transcendental, arising from reflection; they are needed for pragmatic reasons since nothing else could follow without them.

Scientists normally do not consider nonscientific experience as importantly influencing their scientific method, believing that no subjective influence is capable of much influence on the "objective" approach being followed. But it is from *total* human experience (not only the scientific part) that arise the imagination, intuition, and inspiration which empower brilliant discovery.[10] These cannot be denied; it is as dishonest to suppress intuitions as it is to suppress experience from experimental data.[11] That would even apply to intuitions of others since the intuitions built into vocabularies of different cultures are not that much different; there are elements for formulating theses which are common to all.

Most scientists are not eminently successful. There is no good explanation for that, especially when it is evident that they possess a wealth of experience and mental abilities for acquiring and using information and for the reasoning critical to scientific success. Scientists who do not become great are usually able to venture acceptable opinions, but few of their hypotheses withstand critical analysis and become important scientific "truths". Factors distinguishing great scientists are imagination, intuition, and inspiration; such traits can lead to great ideas in individuals who may possess minimal experiences and even scant ability to critically analyze their hypotheses.

Science Progresses by Evolution or Creative Revelation

Human beings become creative through the imagination-intuition-inspiration process. To some the result of that process is considered to appear spontaneously and entirely due to chance, developing from the accumulated mental experiences. Biological evolutionary processes are argued to be responsible for this basis of human creativity. If so, why can't today's scientists, with such vast experience, offer far better explanations of our existence than the ideas of the early Greeks, who had little experience?

Where experience is an unacceptable explanation for innovation, the creative process must have a metaphysical basis. Some would describe it as driven by inspiration from God. For believers, the design of God is revealed to certain people at appropriate times. Such revelation is creative and is implemented through the imagination, intuition, and insight that scientists and, for that matter, all people experience to some degree.

Either Critical Analysis or Birth of Wild Ideas

Critical analysis is essential to a theory's verification or falsification. Confirmation requires experimentation. Experimentation is often considered the most critical factor in initiating the Scientific Revolution. Francis Bacon proposed experimentation as a critical step in the inductive process, but there was a flaw in his belief that an enlargement of experience inevitably leads to increased understanding. Bacon thought that the mere collection of massive amounts of information could not help but lead to revelations of hidden knowledge and truth. In reality, scientific method uses experimentation primarily to discriminate between certain possibilities; observations are generally made and data collected for a specific reason.

Experimentation often depends on technology, the development of which is also dependent on imagination, intuition or inspiration. The insight required for certain scientists and skilled artisans to develop technology can also be viewed as creative, like the inspiration leading to a successful hypothesis. Critical analysis suggests that no hypothesis is completely true or reliable; essentially no scientific hypothesis has been able to withstand the test of time without modification. When modification and discrimination through critical analysis is not possible, hypotheses are no more than wild ideas. Modification of a hypothesis is limited by the degree and variety of feedback from observation and experimentation; with experimentation a hypothesis is no longer a wild idea; it operates with a freedom within certain limits.

An understanding of the process of human thinking for developing scientific as well as other ideas is critical for people concerned about truth. Today many conflicting "truths" are expressed on the origins of the universe and of human beings.

Chapter 5

Rethinking Human Origins

People continue to covet knowledge and seek understanding about their existence, believing that the answers will come through achievements in science. The most fundamental questions, concerning our origins, reason for existence, and ultimate fate, remain unanswered, although many believe the question of human origins is beginning to be understood.

The earliest answers to questions of how life came to be and why life exists in such rich variety were supernatural explanations. Now we also have scientific explanations. But modern evolutionary theory has not led people to abandon theological explanations for a supernatural creation of life. In fact, the powerful evolutionary force of *natural selection*, popularized by Charles Darwin and accepted by many as a culmination of the search for a scientific explanation, is seen by some thinkers as very similar to explanations based on the conscious design of an omnipotent creator.[1]

Seek and Find Design

The Darwinian form of evolution is the natural selection of random variations.[2] Charles Darwin described the

history of life as "descent with modification" in preference
to the term "evolution." He attributed the adaptive
changes of the process to natural selection operating over
long periods on small variations appearing in plant and
animal populations. Natural selection means that nature
(that is, the environment) destroys injurious variations and
allows favorable variations to be maintained and passed on
to subsequent generations. Darwin recognized that small
variations did occur but was unable to explain their cause.
The essence of his theory is maintained today in the
"synthetic" theory of evolution: natural selection by the
environment is seen as working on small chromosomal or
genetic variations, randomly produced by mutations and
recombinations of genetic material.

Completely random variations are not a feature of
nature, however. Variation to produce new forms of
biological material is at least partially nonrandom. At the
molecular level of biology it is possible to study the
self-ordering of amino acids to form protein-like materials
called proteinoids.[3] If amino acids self-assembled in no
preferred manner, entirely by chance, the number of
different amino acids (in the twenties) is great enough to
produce a supra-astronomical number of different combinations, the result being different kinds of proteinoid
molecules. If the initial number of amino acids was twelve,
the number of proteinoids possible would have been
10^{300}, an almost unimaginably large number.[4] Yet the
number of combinations found in real proteins is in reality
relatively small, indicating that the self-ordering of amino
acids is limited and not determined by chance, but is
nonrandom.

The limitation on the number of proteins produced is a
"design factor" and it also has the quality of a "creative
force." We are compelled to explain evolutionary processes as natural selection acting not on randomly produced variants but on nonrandom variants, partly because

of constraints observed in the self-sequencing and spontaneous formation of proteinoids.[5] Here we see a "freedom within limits"; the freedom in the self-ordering of amino acids is nonrandom, that is, limited by the constraints of design placed there by a creative force. This picture of *evolution*, as a stepwise, self-organizing process does not need to be rejected on theological grounds as long as we do not claim that the Biblical account of *creation* is of a relatively instantaneous event.

Nonrandom variation dictates and limits the possibilities for evolutionary change; it assumes the active role. Natural selection determines the final outcome by screening which nonrandom variations are to survive. In Darwinian evolution natural selection by the environment plays the active role in dictating the results. The selection is active because it alone determines what biological material survives. Where variation is random, every conceivable form of biological material is produced and all of it would survive without some weeding-out process. The generator of random variation is chance; it must be passive in the activities of evolutionary processes. With such a passive role, random variation has no determining or limiting influence on which products survive.

If all biological life appears and develops because of chance biochemical events, the final outcome is a product of *chance*. The form and even the existence of all life are indeterminate when life results from chance events, not according to some metaphysical design. But is "chance" the only or best possibility? Can selection of different varieties of life be explained by a determinate process?

Variation Is Not by Chance

If, in constrast to Darwinian evolution, the environment is in reality passive, thereby actively contributing nothing of consequence to determine the product, the selection or favored survival of biological material is determined by variation generated with a selected potential for survival. Such selection is predetermined by the generator of variation so that certain inherent properties of life result in selection and perpetuation of certain traits. Such a rationale for survival makes evolution determinate. Suppose that certain proteinoids formed in the primitive environment of some "primordial soup" survived; should we attribute their "selection" to the proteinoid's inherent properties for survival, or to their passive environment? The proteinoid's inherent properties are determined by those of its subunits, amino acids with a constraint to self-order in limited ways.

Determinism implies a general direction or "time's arrow," in evolution. The direction is evident; life demonstrates increases in relatively objective qualities or quantities, usually generalized by calling them "increases in complexity". In a passive (or constant) environment any increasing organizational complexity must be due to matter constraining itself by itself, an intrinsic selective capacity consistent with the generation of nonrandom variations.[6]

Self-organization and achievement of complexity due to random variation is conceptually unacceptable. The constraint on self-assembly of amino acids at a very basic level illustrates that from the very beginning the organization of life is not based on chance. The process of self-constraining in the generation of nonrandom variation is predetermined; successive variations are produced under new constraints imposed by previously produced variations. Thus, evolution is always under the influence of what has already been accomplished. That makes the generator of variation the active agent for evolution to occur. In this view, the products of evolution are not an outcome selected by something else, such as an environment that is completely a product of chance.[7]

Inherent Creative Determinants

The second law of thermodynamics contributes to evolutionary direction.[8] That law describes the conservation of energy essentially by incorporating energy into structure, thereby rationalizing chemical complexification. The second law also suggests the type of biological life that will be most successful in competing for energy sources, most biological activity in both higher and lower forms of life being primarily concerned with procuring energy in one form or another. Complex organizations may be more competent than simple ones. When they are, since natural selection favors the competent, the complexity of nature increases over the course of evolutionary time.

Classical Darwinism rejects any directive principle, including one that increases complexity. With no teleological direction, Darwin's random variation is undirected, without cause, and with no rationale for improving survival. Darwin was open to the possibility of the environment directing change, even for the hereditary assimilation of acquired traits. Evidently, however, he tended to adopt the position that there is no basis for the inheritance of acquired characteristics, with natural selection as the only causative agent in evolutionary change.[9]

Darwinists thus attribute all adaptation and increase in complexity to chance, relying on random variation and natural selection to account for evolutionary processes for biological life. Physicists and cosmologists, on the other hand, tend to recognize certain properties basic to all components of matter, and the laws those components obey, as causes for the appearance of structure. Formation of matter into structures of increasing complexity in events occurring since the "big-bang" creation of the universe would be considered by them to result from "thermodynamic drive." Current cosmological theory describes the creation of energy and matter from nothing, after which their interactions produced structure in subatomic particles which subsequently interacted to form all the elements. Structure developing in this manner from the time of creation represents a dissipation through structuring, an "evolutionary first principle."[10] Structure is made possible by nature's associative forces and a cosmic assymmetry between potential and kinetic forms of energy.

Structures were formed not by chance but within the constraints of a limited freedom that allowed only a few structures out of astronomical possibilities for others to appear. Appearance of only a few specific structures provides evidence for an inherent factor determining what would evolve. An inherent mechanism dictating how basic matter evolves into substances of ever-increasing complexity represents an "evolutionary" process for the basic chemicals of our universe.

More complex chemicals such as amino acids spontaneously appeared from the basic elements. The number of different amino acids appearing was quite limited, considering the number of possibilities. The limited number of amino acids may have proceeded to self-assemble into proteinoids, but in simulation experiments surprisingly few show up although the number of possible combinations is infinitely large. In such models of "spontaneous creation," processes obeying the fundamental laws of physics and chemistry form substances of only a few types, not the unlimited number of other types possible. That seems to remove "chance" as the dominant factor in the creation of complex materials.

If laws of the universe define the properties of the basic structural units of matter, they would indirectly determine the complex structures produced by matter's further self-assembly to become biological life. As an explanation for the evolutionary process, "chance" has come to represent only the scientist's lack of understanding, so that what little is known must be expressed in terms of probability. That has been understood from antiquity, with Democritus believing that "there is no chance; chance is a fiction invented to disguise our ignorance."[11]

Freedom Within Limits

Variation, generating and constraining succeeding variation, confers a freedom within limits on the evolution of living matter. The freedom permits growth and development of complexity; the limits make that development purposeful, giving it a time direction. Freedom without any limits would be a generator of random variations which cannot result in material self-organization leading to increasing complexity. Yet freedom is built into the system: an outside deterministic factor is not necessary to permit or direct everything that happens at each stage.

Freedom *within* limits becomes an inherent determinism directing "evolutionary progress" to produce innovations for increasing freedom and discriminatory behavior (freedom within limits). Examples of such novelties include the central nervous system, the lensed eye, language, and social organization. Gaining behavioral control of the environment becomes the best insurance for continuing evolutionary "success."[12] The highest organisms, human beings, go beyond Darwinian constraints of fitness and natural selection to acquire an ever-increasing freedom constrained only by the limits imposed by discriminating behavior. Freedom *without* limits, with chance as the generator of variation and the producer of an environment that determines selection, offers no reasonable explanation for progressive evolutionary change.

The evolutionary process progressively changes biological material to increase complexity and that correlates with increasing "homeostasis."[13] In its continuing approach to more effective homeostasis, the evolutionary process endows human beings with a certain order of harmony and balance for the functioning of our biological systems. Homeostasis represents a freedom within limits for that function.

An Evolutionary Creativity for Cultural Progress?

Within the framework of culture, an "evolutionary" process offers an analagous potential for developing organizational complexification with the goal of social homeostasis. That process provides a way for people to live in harmony with their total environment, most importantly including fellow human beings. The possibility for harmony in society would be remote with an evolutionary process in which fitness and natural selection were the determinants of progress.

The overall trend to complexity in organic evolution, and with that a progression to more complete homeostasis, is characteristic of a process with *direction* rather than one which is merely a product of chance. An evolutionary process operating by natural selection to enhance fitness can lead to homeostasis only accidentally, with a very low probability of success. To posit a directive principle, however, defies the current beliefs of most Darwinian evolutionists. They maintain that evolution is completely indeterminate, having no guiding or directing force. That principle is also found in the quantum view insisting that uncertainty attends our perception of nature, which is claimed by some to provide a material basis for the evolution of humanity and the freedom of human consciousness.[14]

Defending the "Truth"

Darwinian evolutionary theorists today seem less concerned about opponents denying evolution as a process for creation than about proponents of a directive principle for evolution, which would suggest determinism. Determinism implies guidance by a design or even by God. The Darwinian consensus in the form of the new "synthesis" effectively silenced the major opposition. Now some assert that "whatever new consensus emerges from ongoing research and controversy, it is not likely to require rejection of the basic tenets of Darwinism."[15] It is critical to most Darwinian evolutionists that their dogma of random variation and natural selection be defended, whether the opposition comes from one extreme such as the "scientific creationists" or from fellow evolutionists who have become "errant" by criticizing the dogma.

One such criticism is based on evidence that all phases in the evolution of life were not gradual. "Punctated equilibrium" is said to offer a better explanation of the empirical evidence, pointing to periods of active evolutionary change interrupted by periods where no changes are seen. Parts of Darwinian theory have to be modified in the light of arguments for punctated equilibrium. Stebbins proposes now:

> . . . to have two types of and rates of evolution occurring in the same organism. The first would be slow constant changes occurring randomly at the molecular level, which are not immediately influenced by the changes in the environment. The second are the relatively sudden sporadic changes which occur in the organism's form or function in response to changes in the environment.[16]

The combination appears to give direction to a process whereby the organism can respond to its environment. Another idea heretical to Darwinian evolution is that acquired characteristics can be inherited.

The ardent defense of Darwinian evolution serves only to protect a particular scientific theory from change. Safeguarding Darwinian principles is not essential for technological progress in any of the biological or other natural sciences. Yet Darwinian theory is ruthlessly affirmed to the exclusion of all other ideas. Why? Because any hint of determinism creeping in may lead to the demise of Darwinism. Thus to Darwinian evolutionary theory the inheritance of acquired characteristics is heresy.[17] When evolutionists completely deny the evidence supporting such inheritance, one is reminded of the Middle Ages, when imaginative, creative scientific thinking was stifled.

In general most biological evolutionists would say that nonbiologists have no right to form opinions on evolutionary theory, being ignorant of the facts of genetics, systematics, biogeography, ecology, and other branches of biology.[18] Only when journalists, jurists, writers, and philosophers are educated in Darwinian evolutionary theory are they considered qualified to speak on evolution.

No Freedom for Other Views

With so many "believers" in Darwinism it is not surprising that essentially every field of human endeavor has adopted evolutionary thinking and evolutionary methodology.[19] Also, many would agree that "applying evolutionary principles has greatly enriched many areas of human thought."[20] Narrowly defining those principles, however, actually constrains the freedom essential for the optimal enrichment of human thought. Freedom of thought about evolutionary processes may lead to the concept that they fulfill some purpose, however. Diehard Darwinian evolutionists are interested in constraining that freedom because belief in ultimate cause and purpose is completely unacceptable to them. Hence widespread belief in random variation, natural selection, and indeterminism strongly perseveres.

Any *ultimate* cause behind biological change cannot be identified or proven by empirical evidence; it can be expressed only in theories. An evolutionary explanation for the development of all the richness of this world is at least consistent with an ultimate cause and purpose. Are Darwinian evolutionists open to modifying theories in which indeterminance is the only guiding principle and random variation plus natural selection the only means for generating change? That seems doubtful:

> There is little likelihood that any new discoveries will force a major modification of the basic theoretical framework that was arrived at during the evolutionary synthesis.[21]

Who Is Spokesperson for Humanity?

Darwinian theory would be an innocuous concept of interest to the curious if the use of its ideas were confined to understanding biology. Darwinists, however, have used their theory as a basis for explaining other "evolutions" outside the sphere of biology, such as the development of human society, language, cultural forms, and ethical principles. Many people find it totally unacceptable to use this philosophical basis of biological evolution for evolutionary explanations in the social realm. Such explanations carry the same implications as Darwinian evolution: chance variation plus natural selection. The meaningless, indifferent, value-free processes of biological evolution are thus applied by evolutionists to the cultural aspects of human society.

Some Darwinists seem to consider themselves fully qualified and appropriate commentators on the entire human condition. Where physics was once believed to be the yardstick of science, today biology, especially in its study of human beings, is argued to provide more important methodology and conceptualization.[22] Physics gained a determinism from the mathematical basis Plato gave physical sciences. But Plato's essentialism proved to be incompatible with scientific thinking in the biological sciences; it could not avoid conflict with Darwinian evolution and its indeterminism.[23] Today, Plato's influence has suffered even in the physical sciences because recent concepts of quantum physics have tarnished the luster of the Platonic mathematical determinancy for physics. That gave greater legitimacy to the biological sciences and their indeterminancy for commenting on the human condition.

A Natural Hierarchy of Life; But Whose?

A composite view of all forms of life suggests a natural hierarchical system in which the wide diversity of biological forms can be ordered. A rudimentary hierarchy of life is found in the biblical account of creation. In ancient times scientists such as Aristotle commented on the natural hierarchy and in more recent times it was developed into a more systematic description by Linnaeus. Darwin amplified that description, arguing that evolution is the true explanation of the natural hierarchy; he held that descent with modification of species produced the system as we find it today.[24] Darwin continues to be credited for that idea although it was not original with him.

Darwin supported an evolutionary basis for the biological system by describing its taxa or suborders as standing in relation to one another as ancestors and descendents.[25] The taxa or orders were established by Darwinists because the theory of Darwin demanded them; the taxa are not phenomena; they are "artifacts" created by scientists.[26] The diversities of life forms are such that they could fit into various systems or hierarchies for classification; when one system is selected, therefore, many other equally good possibilities must be discarded. The orders are *created* to fit, rather than being *discovered* to fit. Darwinists have created them to stand in relation to one another as ancestors and descendents. In another description of evolution, systematics, showing life in all of its diversity through a long historical process, a classification can be developed without an attempt to create a story of ancestors and descendents.[27]

Darwin also saw evidence for evolution in the documentation for a geographical dimension to the history of the hierarchy. The geographical dimension has also been created to fit the theory. If the taxa and geographical variables are not phenomena but rather artifacts, the evolutionary theory built on them may be false.

Criteria for classification or *taxonomy* within the hierarchy of all life must explain similarities and differences. Darwin explained similarity by descent from a common ancestor; he explained differences on the basis of natural selection, producing functional adaptation in an unbroken continuum of individuals. Darwin's position lacks correspondence with reality and results in inconsistency; it is based on the existence of only individuals in nature with no provisions for laws determining form, no relations of an internal kind.[28] All similarities and differences depend on cumulative changes within individuals, without any underlying laws to determine them. The basis of taxonomy still remains to be accounted for, however, and any explanation must agree with the logic and harmony found in the natural hierarchy.

Darwinists believe that fossil records provide documentary evidence of evolution in both time and space. But patterns in fossil records provide at best a severely limited view of phenotypic variation in time and space.[29] Despite that, theories have been constructed from fossil discoveries to support the case for evolution of humans from lower primates (or even simple forms of life). Some evolutionists, who may not argue with the conclusion of such studies, believe that the conclusion's theoretical framework is false, thereby making Darwinian stories of the hierarchy "mere artifacts of that framework."[30] How can truth come from a theoretical framework that is false? In spite of large "gaps" of missing evidence that would be necessary to link the phenotypic expressions of diverse forms of life, fossil patterns have been useful in testing hypotheses of evolutionary processes. Although fossil records provide only limited evidence for evolution, acceptance of evolution based on their findings remains strong.

Current theories on biological hierarchy are different ideas on old questions; they add little to the insights of people centuries earlier. Many of the old ideas are intuitive or appear to have a beautiful logic attractive to many scientists. Such a beauty and logic was found in biological hierarchy in the biblical creation story, by Aristotle and other Greeks, and before Darwin by such people such as Linnaeus. There seems to have been a need to believe in hierarchies of life, suggesting a metaphysical rather than a scientific basis for such belief.

All Purpose in Selfish Genes

Darwinian evolution requires adaptation to persist as an inherent necessity. Perpetuation of desired traits is implemented by the transfer of genetic information found in the germ plasm. That "immortal" information, segregated from the remainder of an organism, is the hereditary determinant and generator of all development and maintenance of life. No empirical evidence exists for the generation of biological form by gene products, however.[31] Moreover, the amount of information needed to control the multiplicity of mechanisms entailed in development through gene expression is astronomical; such an amount can hardly be imagined to be contained in the genome.[32] Also, how can transfer of genetic information be the important determinant of biological form when, in the case of humans, half the genome is "nonsense and repetition"?[33]

Observed differences in life forms cannot be explained in causal terms by differences identified in hereditary factors. Darwinian theory is also unable to explain the observed regularities of form and of generative processes over large taxonomic groups. Evolution is more than a consideration about differences between groups of organisms; it must also explain the regularities seen within taxonomic groups.[34]

If all details of form were determined by information passed in genes which could be freely modified, the result should be an appearance of randomly-variable and infinitely-modifiable forms. With such a mechanism for freely modifying the blueprint in a very random manner, the regularity of form seen in long lineages over long periods of time becomes accidental and improbable. It is difficult to explain how regularity in and constraint on changes in biological form could be compatible with a system of life built on random, accidental variation.[35] Constraint on form to produce regularity is evidence for rules or laws of change operating within organisms.

An alternative explanation of biological development is based on a historical unfolding of a blueprint dictated by genetic information; development is the unfolding of an unconstrained biological potential. For Darwinists the expression of form determined by genes makes the genetic information directing a survival of the fittest an expression of "selfish genes."[35] The selfish genes exist for only one purpose, to reproduce their own kind; in essence life exists as a byproduct of genes which fight for survival through reproduction. Defining genes as "selfish" is a way of saying that they exhibit *purpose*; hence this form of evolution is deterministic, directed by an underlying purpose. The results are not merely a matter of chance.

With random variation and natural selection describing

a process for the development and survival of the most "fit" or best adapted, new problems are posed when "selfish genes" are suggested as the best explanation for that to be achieved. One problem appeared when, with great imagination and a leap of faith, scientists proposed that evolutionary adaptation is no longer concerned with survival of individuals but with survival of societies. A society's genetic makeup must lose some of its selfish genes so that individuals in the group can acquire a genetically determined "inclusive fitness principle," causing each individual to work for the improvement of society as a whole. How could such altruistic genetic activity be compatible with the activity of selfish genes? According to Ernst Mayr:

> Haldane pointed out that an altruistic trait would be favored by natural selection if the beneficiary was sufficiently closely related, so that his survival benefitted the genes which he shared with the altruist.[36]

Evolution is thus able to maintain the concept of selfish genes because the altruism is in essence for the benefit of selected genes. Such a principle would deny humanity any freedom to develop or to choose to improve society; that would all be determined by genes. Where freedom is unimportant because inclusive fitness genes direct evolutionary progress, human beings return to a determinate existence, no longer subject to chance variation and natural selection.

A metaphysical explanation is needed to describe how a process of random variation and natural selection could produce such a phenomenon as the inclusive fitness principle. Humans, supposedly the most advanced example of this principle, show little evidence of a genetically determined drive to work for the improvement of society as a whole. To assert that people are innately good, that they are born that way or have made evolutionary progress to some higher good, is to believe something contrary to human experience.

More Than Selection, An Environmental Assistance

There is now evidence that the environment can induce changes in form and that such changes can be inherited. The possibility of environmentally determined genetic change allows for something other than "selfish genes" to determine biological structure.

With a genetic apparatus lacking any modifier of the "blueprints," biological structure is merely the result of a transcription of the genetic code. All determination of structure depends on that code and nothing changes the code except randomly-appearing mutations. Such a process does not allow organisms to actively modify their structure and function in order to gain an advantage over limitations imposed by the environment.

Biological organisms have properties collectively called regulative behavior for purposively responding to the environment.[37] Regulative behavior directs development over and above that dictated by a readout of the genetic code. This behavior establishes a completely dependent interrelationship between each point in a developing organism's field of growth so that the state of each and every part determines and is determined by the state of each adjacent part. Structure and function in one part are dictated by conditions in adjacent regions. Following disturbance at some point or a disordering among more than one part, regulative behavior restores the normal relational order. Living matter thereby exhibits "field properties."

In the *field* theory, the life-development process is determined both by inherited factors in genes and by environmental factors, which combine with certain universal factors to generate specific form.[38] The universal factors include a continuity of cytoplasm that is guaranteed from one generation to the next. The molecular composition of the cytoplasm perpetuated in descendants is maintained by the activity of genes which are reproduced as part of the living state. The cytoplasmic composition is also influenced by environmental factors. In this theory the products of gene activity influence the organization and expression of field conditions by defining parameters of the cytoplasm and by affecting boundary conditions within the cytoplasm. This is in addition to the well-recognized property of genes for contributing to the invariance of a living state.

The field theory represents a basis for unification in biology such as that sought for explanation of forces and particles in physics. The unification takes into account certain universal properties of the living state (such as regulative behavior to maintain normal relational order) and the particular cytoplasmic properties of each individual field of the living state; those properties are determined by specific gene activity plus the influence of nongenetic, environmental factors. Expression in the living state is by a system of activities blending those of the universal and the particular, according to this "unified field theory."

Descriptions of all living states using field theory provides an analytical basis for a rational taxonomy, a systematic classification of morphologies in terms of relationships between field solutions. The field theory goes beyond depiction of the living state as merely the historical unfolding of biological potentials contained in genes. It offers an alternative to the idea that expression of all parameters of the living state is dictated by the activity of genes.

There is More to Life than Genes

The hereditary apparatus for expression of an organism's development must include oocyte cytoplasmic factors as well as the nuclear genes contributed from both parents.[39] The cytoplasm of the oocyte is both a carrier of heredity *determining* development and a channel of communication for the environment to *influence* development and possibly modify the evolutionary process. Natural selection does not provide a means for the environment to direct an adaptive evolutionary process, because it provides no means for registering experience and assimilating it for future generations.

Development in any organism is not the result of a program in which all results are merely the consequence of translation of nucleotide (DNA) sequences in the genetic code, with code variations appearing spontaneously or accidentally. The genetic code does not express itself. The cytoplasm controls genetic expression by dictating which sequence of DNA will be exposed and copied and by manipulating DNA sequences through cutting, splicing, transposing, rearranging and amplifying genetic strands.[40]

Strands of DNA in chromosomes are covered by proteins. The DNA sequences must be uncovered for manipulation or copying into a different nucleotide sequence (RNA) for directing protein synthesis. Nothing happens to accomplish that until factors in the cytoplasm uncover the genes and direct which genes will be copied. Thus, the cytoplasm is the director or controller of all expression of the genetic code.

The genetic apparatus is not stable but extremely fluid and subject to modification. Its potential for instability can hardly be reconciled with a process of development that proves to be stable and reliable. The fidelity of the developmental process in a living state where genetic translation has a potential for great fluidity must be at least partly explained by the physiochemical reactions in the environment of the developing system as well as by the assimilated experiences of the cytoplasm and nuclear material.[40]

In contrast to the consequences of chance genetic variation plus natural selection by the external environment, assimilated experiences operate to prepare an organism for anticipating the environment to be experienced. Thus the living state has become active rather than passive in the selection. Organisms develop and exhibit adaptive evolution by relying on outside information or cues from the environment, so that the organism adapts and the environment serves its needs.

Adaptation: Increasing Freedom Within Limits

In this newer view of evolution, human beings are more than merely products of chance, more than the result of environmental selection of random variations. Subjection to such selection denies humans any freedom. But ability to develop by an adaptive process for actively modifying their biological state gives humans a freedom within limits. The freedom is permitted by the possibility for assimilation of experience and its transmission to subsequent generations. The limits are imposed by constraints in the nuclear genetic code and in the cytoplasmic determinants that ensure an invariability for structure and function. Such freedom within limits offers the possibility for active assimilation of experience, for testing, and for modification of behavior so that a greater homeostasis of the living state with the environment can be achieved.

The Environment Induces Adaptation

Environmentally induced adaptations shown to be transmitted to one or more successive generations support the idea that experience is assimilated into the hereditary apparatus.[41] For example, immune-mediated mechanisms experimentally stimulated to cause eye defects in male rabbits are passed on and expressed in a progressively worsening lesion for at least nine generations. Male mice made tolerant to the foreign major histocompatibility antigens transmit the trait to their offspring. Treatment of animals with an experimental drug to produce a form of mild diabetes leads to progeny with the problem through at least three generations. Alterations produced by LSD in insects passed through the male for at least two generations. Thus some environmental modifications of the genetic apparatus are passed to offspring, in at least some cases by a transfer of genetic material from the somatic to the germ cells.

Until recently no one could see how perpetuation of environmentally induced adaptations could occur. Now it is known that the transfer of genetic information is possible by naturally occurring endogenous retroviruses, which are found in all cells and transmitted vertically (in utero) from one generation to the next.[42] The normal role of intracellular retroviruses may be to provide a means of intercellular communication. Such retroviruses would be capable of capturing mutant somatic information (changed by environmental influences) in the form of RNA, the molecules that direct protein synthesis.[43] In its normal synthesis RNA is copied from DNA in a cell's nuclear material. In this unusual case the reverse process would take place, with the information in RNA being copied and stored in DNA in a cell's genetic information. How could that happen? The captured material would be acted on by endogenous retroviruses, using an enzyme they possess called reverse transcriptase (RNA-dependent DNA-polymerase), to copy the RNA information into DNA, making it part of the genetically transmitted information. Mutant or acquired somatic information is common to the animal exposed to unusual antigens and producing a large amount of antibodies in response. That immune response is directed by RNA and now retroviruses explain how that response can become part of the "memory" passed on to progeny through the germ plasm.

In addition to an intercellular movement providing a form of communication implemented by movable genetic elements, it is also possible that genetic information may cross *species* barriers. Such transposition might occur through a viral vector. One piece of evidence for that idea is the structural similarity of the protein leghemoglobin, found only in legumes, to the globulin protein of vertebrates.[44] The similarity is not explained by a common evolutionary origin and probably indicates gene transfer.

An evolutionary scheme for the development of mammals from lower animal forms of life is said to be supported by the close similarities in some peptide regulatory hormones found in mammals and insects. The possibility of interspecies transfer of genetic material introduces an alternative to descent by evolution.

Seeking the Source of Human Uniqueness

Although many ideas are held regarding the origin of life, most people consider one or another form of evolution plausible. Belief in evolution is maintained despite certain deficiencies and huge gaps in understanding for any of its theories. If all aspects of evolution purporting to explain the progressive development of life from single cells to the appearance of human beings were accepted, the problem of explaining an evolutionary development for the human mind would still remain. That problem looms importantly in the age-old question: Where did I come from?

Seeking the Missing Link

Can evolutionists adequately explain the emergence of human mental potentialities from the mental processes of subhuman primates considered to be most closely related? The "missing links" bridging the gap between subhuman primates and humans are identified as very old skeletal remains showing progressive and continuous changes toward *Homo sapiens*. It is not clear how such intermediate skeletal forms would substantiate a claim for the emergence of human mental processes from lower forms of animals.

The skeletal remains of primitive humanoids show relatively minor differences from those of modern human beings, with one exception. Examination of the skulls shows that the brains of some of these primitive forms were only one-third the size of human brains. It is generally assumed that the evolutionary process resulted in an increase in human brain size that was remarkably rapid.[45] It seems somewhat paradoxical that the most complex and least understood organization of living matter should evolve so rapidly in comparison to a slow evolution for relatively simple structures common to life.

The small humanoid ape brain of subhuman primates is structurally and functionally similar to the human brain. If evolution was responsible for the changes, the change in size was more remarkable than other changes. The relation of brain size to sophistication of function is not well understood; some mammals with the body size of humans have larger brains. "Brain size cannot be taken unconditionally as an indicator of human intelligence; a hominid with a bigger brain case is not necessarily smarter."[46]

A Biological Basis for Human Potentiality?

Mind and reason are of foremost interest in the study of human development. One approach to understanding the human mind has been to study biological activity of the brain at cellular and neurochemical levels. The philosophy behind that approach regards the mind as essentially a product of neural processes. Such a reductionist approach has shed no light on how human beings reason, a level of operation above the innate, instinctual reactions common to all life. The general success of reductionism in providing technological achievements through science leads people to believe that science will eventually account for the mental potentiality unique to humans.

The approach most commonly used for understanding the mind has been antireductionist, however, making it thereby seem, less scientific, less objective. Antireductionist approaches include the philosophy and methods of traditional experimental psychology; the idea that the human brain is capable of inborn, spontaneous thought or innate ideas; study of processes basic to computer operation; and frequently a naturalism holding that the natural world is the whole of reality. In such a naturalistic philosophy, there is no supernatural or spiritual creation, value, control, or significance, and scientific laws are capable of explaining all phenomena.[47]

Antireductionist studies of human behavior are designed to describe structures and processes essential to the operations of reasoning and to the manipulation of symbolic information.[48] Processes for encoding, storage, retrieval, transformation, and transmission of information have been studied, on the premise that behavior is rule-governed and generative. The fact that transformational rules intervene during encoding processes makes goal-directed problem solving possible. For many scientists the mind is merely a huge computer; yet even a thorough knowledge of all computer capabilities and possibilities would not begin to account for the human potentiality.

How Does the Innate Become Innate?

The human brain is unique in that it alone is associated with mind and reason. The belief that ideas and thoughts can be innate or spontaneous suggests a determinism which most scientists are unable to explain or accept as such. Spontaneity indicates that something appears through no activity on the part of the human will. If thought cannot be explained by chemical events at neural processes, and if it is unique to human beings, is inborn and innate, and operates according to rules or laws, then some deterministic factor seems to be at work.

There is no evidence that other animals, including other primates, think abstractly and generate ideas. No adequate evolutionary explanation is given for the acquisition of innate human ideas and thought by only one form of life; nor can scientific evidence establish the naturalism that is proposed to dictate human development. By definition there is no supernatural or spiritual warrant for such naturalism, making its rules and laws simply "there." A naturalistic science cannot explain why only humans have this mental ability or what constitutes its scientific basis. No form of science can yet do that.

Humans are unique in their language and methods of communication. Norman Chomsky states that humans are innately equipped with a pre-programmed "language acquisition service." This innate attribute is suggested to specify universal constraints on grammars and to permit "a child to transduce the grammar of his native language from the 'degenerative' input of the actual utterances of other speakers."[49] That innateness appears to be inherited with the human potential, much as an instinct is inherited in the repertoire of animal behavior.

Language is not *simply* inherited, however, since humans isolated from other humans never show any ability for language.[50] Attempts to teach people to speak after they have matured are largely failures. Instead of humans being innately equipped with a pre-programmed language aquisition service, they seem to have been given a potentiality that includes an ability for communication by way of lan-

guage. That ability is not expressed unless language acquisition is an important part of a child's development. Human
beings are thus innately equipped with only a potentiality
for communication, which requires development.

Selected to Construct an Environment

People think abstractly and create their own environment both in thought and in real surroundings. Darwinism, however, does not allow organisms to mold their
environment but requires living forms to be selected by
their environment. It is difficult to see how an abstractly
thinking organism could be selected by any environment.
If Darwinism is correct, how did we escape selection and
learn to create our environment?

Some believe that human beings are an aberration which
is alien to this cosmos.[51] Born with an immanent deficiency,
we lack pre-given structures and regulatory processes for
fitting in and being a part of the cosmos; we appear with
a poverty of instincts. If we have "fallen back" out of the
ordered arrangements of nature, how then did we survive?
Humans overcame this deficiency by developing substitutes
for the actions of regulatory processes or instincts. Rhetoric
as the essential substitute enables us to overcome our critical
deficiency. We were fortunate to escape selection by the
environment, for it was only by chance, and very improbable at that, that the human creature appearing with a
poverty of instincts should have survived.

Humans with the ability to think abstractly, and all other
life without that ability, both selectively construct the
environment in which they live. To different degrees,
species of organisms are, as emphasized by ecologists,
active in shaping their niche as well as being shaped by it.[52]
This allows for a "selective constructivism": an organism's
innate algorithm allows it to be selective in using input
from the environment to construct its most favorable

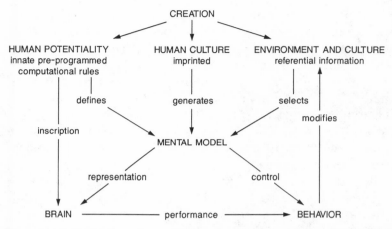

Figure 2. Behavior based on selective constructivism. Behavior is based on limits and freedom; humans have the ability to select from the environment to modify the mental model and behavior which it directly or indirectly controls. Creation is the ultimate basis, inherent in the possibility for evolutionary progress through cultural input and the choice of selection of referential information.

niche.[53] Such a process applies both to immature development and to function at maturity.

Darwinian natural selection and classical conditioning and learning are best explained by assigning a primary role to behavior for any increase in adaptation by the organism.[54] In this paradigm the environment plays the effecting role by directing which of the organism's innate algorithms or computational rules are to be selected for survival.

In selective constructivism an organism's computational rules are not derived or selected by the environment (Figure 2). They are innate rules enabling the subject to select environmental inputs relevant to the organism's process of constructing its own internal and external environment. On assimilation, that which is selected can possibly modify computational rules, further modifying the selection of any future input from the environment.

The innate computational rules represent the human potentiality. It is evident that a unique kind of behavior cannot be expected without any input from elsewhere.

The model in Figure 2 is modified from that of Chris Sinha.[55] To describe the human potentiality merely as innate is to assign it to the metaphysical realm, beyond scientific understanding. Here it is represented as a product of creation or special design. Both Sinha's model and the one here are profoundly anti-Darwinian in their implications, amounting to arguments in favor of special design. The model assigns to humans a freedom within limits with a possibility for enlarging those limits either through modification of the potentiality's computational rules or through human culture, both being directed as part of a special design.

Selective constructivism enables the organism or subject to select its environment, whether that environment exists as an abstract mental model or as the world outside. Human beings create abstractions and actual sensory perceptions of the world in their artifacts, institutions, practices, symbols, utterances, and languages—all being a part of *culture*, which is assimilated anew by each succeeding generation. They are not created by the computational rules constituting the human potentiality; that potentiality by itself (evident in isolated humans) results in no appearance of humanity's cultural manifestations (its mental models).

The computational rules define (grant unique freedoms) and limit the extent of expression of the human potentiality; the cultural input imprints individuals with the generator of the potentiality's expression (confers freedom). The cultural imprint also limits the expression of the mental model; at the same time it introduces a freedom for testing and expanding the limits. People are freely able to select referential information from the environment for determining behavioral responses. The *selection* does not limit their behavior. In "natural selection" behavior is selected by the environment; in this model behavior is subordinate to the computational rules creating and modifying input to an abstract mental model, thereby dictating and modifying behavioral response. Assimilation of environmental input and adaptation constitute a mechanism for feedback and modification, a mechanism totally inconsistent with Darwinian evolution.

In the cultural imprint, internalized during human development, rules can be socially negotiated. Intersubjective agreement allows such rules to become human laws rather than accepted as metaphysical dictums. Human laws can be based on a philosophical naturalism in which the natural world represents the whole of reality and scientific laws are able to explain all phenomena. In a naturalistic scheme, overall systemic development is driven by endogenous processes, controlled by each individual. In such a scheme, humans beings are seen as controlling their own destiny and having the potential for answering all questions and solving all problems.

If assimilation and adaptation (or accommodation) is not regarded as decided by mutual agreements among human beings, it would have to be represented as derived from timeless formal abstractions, innate laws, or laws from a theological perspective—i.e., laws of God. But it is no longer fashionable to believe that the imprints on human beings by computational rules and culture are timeless formal laws designed and planted in humans by God.

Neither a philosophical reductionist nor an antireductionist is able to produce scientific evidence for explaining the potentiality innate to the human mental process. Antireductionists grant human mentality the possibility of innate ideas and spontaneous thought in addition to computer-like operations, in a system based totally on a naturalism where scientific laws explain all phenomena. To acknowledge innate rules or thoughts and a spontaneity contradicting rule-governed behavior, however, is inconsistent with scientific reductionists' attempts to provide a causal explanation for appearance of the human mental potentiality. Innateness grants a deterministic metaphysical basis, implying a failure of scientific laws to explain all phenomena of the mind. Innateness also defines limits for expression of the potentiality, whereas spontaneity grants that expression a freedom.

Progress by Enlarging the Freedom with Limits

An evolutionary process could work on the human potential to enlarge its freedom and gradually allow human behavior to reduce the necessity for limits. Enlargement of human freedom within limits has been seen throughout history. People were once severely constrained by laws such as the Mosaic code which specified every aspect of life, or by superstitions that guided their total activities. Constraints have been markedly reduced as the environment has become more habitable and knowable.

Scientific knowledge is largely credited for driving the human potentiality and claiming it to approach perfection. Actually, however, people have interfered with any growth of the human potentiality, as illustrated by twentieth-century warfare with its holocausts paling all previous wars. So instead of progressing toward ultimate knowledge and understanding or the technological mastery that would eliminate needless deprivation and suffering, people have generally failed to control their destiny. Whatever people have learned about their origins has not aided their attempts to improve the human potentiality.

A Universe with Goal-Directed Processes?

Goal-directed processes exist throughout nature and Darwinian evolutionists believe that they develop through a process of random variation and natural selection. We have already discussed the self-assembly of atoms and molecules at the very beginning of life's evolution, where variation cannot be said to be either random or due to chance. The molecular basis for all life is subject to constraint by physical laws. Do those laws (of gravity, thermodynamics, etc.) themselves represent a world of coincidence and chance or a world that exists as a consequence of purpose?

Scientists and philosophers assume that every phenomenon must have a cause. Yet the idea of a purpose stemming from some final or ultimate cause behind the universe itself is rejected without reservation by many scientists. Scientists do address the question of causes for the physical laws of the universe, however. One would think that chance processes, such as the random variation and natural selection allegedly providing the basis for evolution, would also be invoked for the appearance of physical laws and properties of elements essential for the formation of the first living matter. For example, the force of gravitation with its specific interactive strength might be said to appear because of chance and not for any reason. It is not unreasonable to expect scientists to give either a chance or a purposeful basis for the physical laws.

Scientists sometimes formulate theories in one field of science and consider them as special cases of a theory or law from some other branch of science, or as a specific example of some underlying universal principle. The indeterminism that arose from quantum physics has been used to theorize on the human condition, with some concluding that people lead a meaningless existence in an indifferent universe. For others that indeterminism supports a lack of meaning for all living matter and its evolution.

Only Order Is Possible

The existence of physical laws precisely defining certain properties of matter and excluding others suggests that the evolution of matter and of life is an inevitable process, no other possibilities being open.[56] Any degree of "random variation" actually seen on an evolutionary theme reflects an indeterminancy, a certain freedom within limits imposed by the narrowly defined physical laws. The constraining limits represent the beauty, order, and grandeur expected in a purposeful design. A freedom within limits grants a degree of creativity to a process that would be stifled by a completely deterministic process.

The Darwinian evolutionary processes of random variation and natural selection have been claimed to provide a mechanism for creation in a meaningless and indifferent universe.[57] Those two processes offer no means for implementing creativity, however. Random variation generates every conceivable possibility; creativity relies on discrimination with regard to what is produced. In reality the variation in biological evolution is nonrandom because it is limited by certain restrictions, as we have seen in the limited number of different amino acids generated from the beginning. The pre-life evolution of matter itself was also severely restricted, with only a relatively few types of molecules arising from self-assembly of smaller ones. The natural world is deterministic, with the innate properties of the basic subunits being predetermined.

The first step of Darwinian evolution cannot be creative if it is completely random. To be creative, Darwinism must rely on natural selection, the final step in the overall process. It seems more appropriate to credit the creativity to the generating factors, however, rather than to the step that sorts and culls. Biological evolution should indeed be recognized as a creative process and a discriminatory nonrandom variation should be given credit for that creativity.

At levels of life with greater complexity, the evolutionary products found at each stage are a creative compromise, since it is impossible to improve simultaneously all components of the phenotype to the same degree.[58] The compromise results from a continuing restriction, in this case on the genes and cytoplasm that are transferred to offspring. The creative force is a combination of the creative freedom of nonrandom variation plus restriction by innate properties outlining the limits within which development is free to express structure and function.

Is Order by Chance or Design?

The problems of attributing purpose or purposelessness to the rich lore of living matter on earth began at least twenty-five centuries ago. The distinction between chance and purpose as an explanation for adaptation in biological matter was recognized by Aristotle. Aristotle's problem of selecting either chance or necessity to explain the diversity of living matter was reexamined in the seventeenth century. In "natural theology," the wisdom and design of God was seen as manifested in all creation, eliminating any conflict between science and theology. For the natural theologian, nature was "convincing proof for the existence of a supreme being, for how else could one explain the harmony and purposiveness of the creation?"[59]

The same scientific evidence used to support arguments for natural theology's design was subsequently used to develop *evolutionary* theory.[60] The new theory began to replace the old, partly because people became uncomfortable with the inconsistency of an imperfect world being guided by the hand of an omnipotent creator. Evolutionary theory developed before Darwin, lacking the rich experience of observation which he was able to accumulate.

With the development of Darwinian evolution, Aristotle's problem of whether biological changes were due to chance or necessity continued as before. With no new critical empirical evidence, reformulating the scientific concepts could provide no better understanding of evolution or its causes. Real progress in conceptualization should lead to answers to the most basic questions on human origins. It could hardly be called progress for people to be told that their existence is meaningless in an indifferent universe. That is not an *answer* to the questions but rather an indication that, by their own efforts, people have found no answers. Although progress can be claimed for technology, to render human existence meaningless is poor evidence of progress in theory.

A Beauty and Satisfaction in Order; So Unify

The unsatisfactory state of evolutionary theories on diversity and adaptation of living matter has continued, although Darwinians claim that new concepts have led to a unification of evolutionary theory similar to that developed for physics. The unified theory, or evolutionary synthesis, is considered today's biological paradigm.[61]

The synthetic theory affirms Darwinism with greater conviction than before, rejecting any form of conflicting concepts such as those based on essentialism, inheritance of acquired characters, internal perfecting or finalistic principles, or any evolutionary or creative process that is not gradual.[62] The neo-Darwinists accept no form of these concepts, vehemently rejecting any suggestions that they might bear some merit. Thus any trace of determinism or design is prevented from infiltrating evolutionary theory.

The "synthesis" brings together empirical evidence from the study of natural history and newly acquired knowledge from experimental genetics. Yet "Evolution, as we now realize, can be inferred only by indirect evidence, supplied by natural history, ..."[63] Experimental genetics is limited to describing changes that can produce only minor evolutionary adaptations. Thus today's Darwinian evolutionary dogma combines genetic understanding of minor changes with the indirect evidence from natural history to formulate "wild but educated guesses."

What would be considered significant progress in biological science? Progress for its technology is unquestioned. But for some, progress consists principally in development of scientific *concepts*, with new or radically transformed concepts considered often more important than the discovery of new facts.[64] That is unacceptable since today's evolutionary theories are severely "underdetermined" (many theories explain a body of evidence as well as any one theory). Today's evolutionary synthesis may well be viewed as a myth by tomorrow's scientists, in much the same way that scientists of today regard the concepts of the past. As time erodes the legitimacy of scientific theories they evolve into tomorrow's myths.

What Rethinking Can Do

In summary, the most important evolutionary progression for human beings would entail relationships with other humans rather than merely biological transformation. If reason and science could enable people to improve themselves and the lot of humanity, improving understanding and solving human problems, human beings would be directing the most significant part of their own evolution. For Darwinian evolutionists, reason (believing it to be the capacity humans have for discovering truths) has survival value and evolved in a similar way as for our biological form. But, regardless of how reason appeared, this is an "empty formula" because humans may have utterly irrational beliefs that enable survival and could have most rational beliefs that would quickly lead to extinction.[65] If reason enabled survival it would not depend on knowledge of human origins since that has hardly improved the human condition. It remains to be seen whether any of the newer evolutionary concepts will ever be successful in providing the knowledge sought by Adam in the Garden of Eden.

Chapter 6

Reason's Golden Age

Bertrand Russell once articulated human faith in science by saying that "what science cannot tell us mankind cannot know." Scientists continue to believe in an *order*, in what might be called a "master plan" consisting of universal laws for the universe. A few centuries earlier that view took the form of a determinate universe, with the determinism likened to a great clockwork set in motion by a divine hand at the beginning of time and then left to run undisturbed. In such a scheme, if the master plan could be understood, it would be possible to predict accurately all happenings; nothing would be left to chance. The idea appeared to be supported by Isaac Newton's description of the laws of motion and of gravity, which were successfully applied to describe the detailed motion of the moon and planets. Scientific observations have supported the idea of a determinate universe operated by fixed laws. That view began to change in 1905 with the work of Albert Einstein.

A New Age Begins

Einstein's new concepts transformed classical physics into something radically different. They essentially created a new branch of physics called *relativity* and resulted in the creation of another new branch called *quantum mechanics*.[1] The first of these innovations did not undermine a deterministic basis for the universe but the second did. Einstein also confirmed the atomic theory of matter, demonstrating the validity of ideas proposed by early Greek scientist-philosophers.

Progress by Unification

Einstein was a great scientist but not because he discovered new phenomena; he is recognized for unifying ideas and observations into new concepts which were then confirmed with mathematical explanations.[2] For example, he is credited with the proposition that the time recorded by an apparently moving clock will be less than that of an identical clock apparently at rest. That phenomenon of "time dilation" suggests that a person traveling at high speeds in a spacecraft will age more slowly than that person's identical twin remaining on earth. The concept of time dilation can, in retrospect, serve as an interpretation of the results of an 1887 experiment attempting to measure the earth's velocity relative to its atmosphere; the old observations can be fitted into the new, more difficult concept.[3] Einstein also proposed that gravity has an effect on the time a clock records; the apparent time recorded slows when the time-recording device is in a field where the force of gravity is increased.[4]

Great efforts have been made to prove Einstein's theories of the effects of motion and gravity on time, and many believe that considerable evidence has been found for their support. Philosophically (or on the basis of "common sense") the concepts have been difficult for most people to accept, even though they can be explained mathematically.

Disturbing the Clock

Modern methods to prove time dilation are based on observing differences between radioactive decay rates of mesons at rest and mesons moving at high speeds in a circular path.[3] Differences in decay are attributed to differences in speed but might be due to differences in gravitational forces on subatomic particles escaping in the decay process. Gravitational force has effects on the subatomic structure of other substances; when neon is accelerated in an atomic beam in experiments to demonstrate time dilation, changes in subatomic structure are manifested by transition frequency differences between defined electronic excitation levels.[3] The results of such experiments come from indirect observations. It is unknown to what extent the results are influenced by the instruments used and thus to what extent they represent an observer-created reality rather than true reality. A similar effect may explain the influence of the earth's gravity on time: the "slowing" of time by increasing gravity must be measured by man-made devices.

All such experiments have in common the movement or a change in position of a time-recording device; the movement in a gravitational field alters the effects of gravity on the device's structural particles. Gravity does have effects on the structure and operation of material systems, including living systems.[5] If gravitation affects the structure of molecules associated with physiological functions, it should not be so surprising that it can affect the properties of even subatomic particles, in living matter or in a time-recording device.

Progress by Confirmation of Greek Thought

One concept of the early Greek scientists, the idea that all things are in perpetual motion, was confirmed in the early twentieth century. In Aristotle's concept of space and time, without motion there is no time; the motionless is out of this universe.[6] Time can be measured only by some change taking place, so time "occurs" only where something is changing. Change requires motion, and since all things are in perpetual motion they are "in time" as well as in the universe. If acceleration of a meson towards the speed of light slows its radioactive decay and perhaps its perpetual motion (that is, its "random" motion) so that there is less time, as it approaches "no time," it would appear to be outside of this universe. With no movement there is no passage of time; according to Aristotle it is outside of this universe. Only God who is outside the universe is unchanging and therefore motionless and outside of time.

Einstein confirmed the atomic theories of many centuries earlier by using observations of the previous century on Brownian motion and relating it to the kinetic theory of gases developed in the 1870s.[7] Einstein's contributions included an equation to measure accurately the mass of the atom and an equation describing the diffusion of a suspended Brownian particle through a fluid medium, both of which led to the establishment of statistical mechanics. Recognizing the difficulty or impossibility of measuring the movement of single particles, scientists have used statistical methods to describe the probability of the behavior of a group of molecules. A tendency to describe the natural world in probability terms rather than as a certainty marked the beginning of determinism's demise.

From a Determinate to an Indeterminate Universe

The success of science is predicated in part on its ability to predict events. The impossibility of predicting the movements of a single particle suggested that its movement was completely random and thus indeterminate. On the other hand, the aggregate movement of a group of randomly moving particles could be predicted with a certain probability. The indeterminate behavior of individual atoms in a solution was not enough to make

Einstein believe that the whole universe was indeterminate.[8] That was left to the later formulators of quantum physics. Although it is not completely accepted by the societies that supported its creation the indeterminate concept of the universe is considered a fruitful intellectual achievement.[9]

The transition from the determinate to an indeterminate view of the universe parallels application of the idea that the whole of the universe is described by mathematics, to be revealed in laws of the "cosmic code." The new indeterminate view of the universe is the mathematical expression of the laws of the universe in quantum physics. The culmination of the transition is expressed in one writer's belief that "mathematics, with the demise of theology, has seemed to be the bastion of *a priori* knowledge."[10]

Before pursuing this further, we should note that some believe that "the fundamental laws of physics are not true" (fundamental laws include the Maxwell and Schrödinger equations, and general relativity) "but they are explanatory."[11] The phenomenological laws of physics, describing specific phenomena such as superconductivity, however, are (or tend to be) regarded as true.[11] Perhaps mathematical physics should not be taken too seriously, even though it may have become the bastion of *a priori* knowledge. Physics (including quantum physics) tries to explain all phenomena of the physical universe, using mathematics as a means to that end, but it is *not* the investigation of a particular mathematical structure for the world.[11] One's view of the role of mathematics in structuring the universe influences one's decision about whether the universe is determinate or indeterminate.

Indeterminism implies that physical events are forever unknowable and ultimately unpredictable. Beyond mere human inability to know where an atom will move in its perpetual motion or when a radioactive atom will decay, the indeterminist believes such matters cannot be known even in the perfect mind of God. If there is a God or grand design in the minds of physicists, it is a cosmic order with a mathematical basis for the order's expression.

A Degree of Uncertainty with a Touch of Weirdness

Indeterminism attracted many disciples in the early twentieth century, after physicist Werner Heisenberg emphasized what atoms do, not what they are. Heisenberg developed new mathematical methods for describing atomic processes and for expressing physical intuitions in precise mathematical terms, leading to development of his "uncertainty principle."[12]

Along with the work of another contemporary physicist, Niels Bohr, whose mathematical work led to the "complementarity principle," the uncertainty principle convinced scientists of the correctness of the new quantum theory. The work of Heisenberg and Bohr revealed an internal consistency for the quantum theory, a consistency requiring people to accept a new concept, the basic tenets of which necessitated renouncing determinism and objectivity for the natural world. New mathematical equations described the universe as indeterminate. Those equations raised certain conceptual questions that led physicists to coin the phrase "quantum weirdness." That phrase seems appropriate for quantum theory's mathematical descriptions of the microworld, which were inconsistent with intuition, common sense, and certain physical features of the macroworld.[13]

If mathematical expressions are accepted as our most reliable means for describing and understanding the universe, it is indeed puzzling that they should lead to such a situation. Using terms like "crazy" or "weird" to describe quantum theory seems to be a more or less metaphysical way of glossing over a major scientific inconsistency.

The indeterminism of quantum theory requires probability to be causally determined into the future, but not individual events, because patterns or symmetry could not be found with which to predict individual events.[14] The "randomness" observed does not rule out the possibility of patterns or symmetry, however. It is possible that (a) the observations are accurate but insufficient to describe the symmetry or (b) the act of observing changes the thing observed so that the observation can never reveal any inherent pattern, making any predictions impossible. When events appear to be random, some people conclude that the universe is indeterminate; others conclude that there are limitations to the human ability to observe, know, and understand. When the natural world appears to be predictable the universe becomes determinate and guided by God, a "grand design," or a "cosmic code."

Is Progress Measured by New Concepts?

The Heisenberg uncertainty principle says that for a particle in motion it is not possible to establish simultaneously both its position and its momentum, momentum being a function of the particle's mass and velocity.[15] Quantum uncertainty has been acclaimed as "one of the great scientific ideas, not only of the twentieth century, but in the history of science."[16] Recall that to Aristotle it made no sense to attribute motion to a body at an instant.[17] In essence that says that a body's position at an instant can be described, but that observation does not allow any conclusions on velocity and thus momentum. We have already cited Heisenberg's familiarity with Greek thought.[18] We have also stressed that the more or less radical transformation of old concepts is often more important than discovery of new facts, a view supported by Ernst Mayr in *The Growth of Biological Thought*.[19] Although the Heisenberg uncertainty principle was received as a new and

revolutionary concept, its similarity to ideas held by
Aristotle suggests that it was not essentially new.

Observing Dimly and Indirectly, as Plato's Shadows

Indeterminism gained strength primarily from the uncertainty principle, conceived out of belief in quantum
theory. Quantum theory was developed from scientific
observations as well as from theoretical mathematics.
None of the scientific observations were made directly on
individual parts of the atom; all were indirect, using
instruments that often required destruction of atoms for
examination of the subatomic components. Investigations
using such methods change atomic substructures, however. Such changes are well documented for methods used
to study electrons. Quantum mechanics thereby becomes
an observer-created reality and its theory a theory of an
instrumentally detected material reality.[20]
The indeterminism of quantum theory is based not on
phenomena directly perceptible by human senses but on
what instruments detect; the nature of the detected depends on the experimental methods and types of instruments used. An objective world appearing the same to
everyone's senses each new day is replaced by an observer-
created world with its appearance determined by how it is
examined and possessing no objective reality. The quantum theorist accepts that appearance as reflecting the
reality of the natural world; such a *quasi* reality depends on
many assumptions, however.

Quantum Leaps of Faith

The quantum theorist has faith in reason (i.e., mathematics) for understanding the world, no matter how

"weird" the mathematics may seem to make it and despite the failure of the human senses to comprehend and accept such weirdness. Because of quantum theory the whole microworld loses objectivity and becomes indeterminate. Since the microworld is the basis of the macroworld, does its indeterminism render the entire universe indeterminate and devoid of any objectivity? When faced with this question the quantum theorist generally explains that the world of electrons and atoms is qualitatively different from the world of tables and chairs, so that "quantum weirdness" does not exist for the macroworld.[21] Such reasoning is not entirely satisfying. More importantly, the macroworld seems to be allowed to retain objectivity but not determinacy. The world as a whole remains an objective reality but paradoxically becomes indeterminate, which seems to represent a logical inconsistency. "Belief" in quantum theory makes it possible for scientists to accept some wild theoretical ideas of scientific weirdness and deny any determinacy for the world.

Quantum theory has introduced some other difficult problems as well. A scientific principle important to physicists is that of local causality: the idea that distant events cannot instantaneously influence local objects without any scientifically explainable mediation.[22] Quantum theory violates local causality, so that a particle in one part of the world can be instantly influenced by another particle or events far away, even half-way around the world. An event on a galaxy hundreds of light years away, even, could instantly influence an event in our world.

Although the logic of instantaneous influence from a distance may be mathematically correct, it has been unacceptable to scientists, generally for philosophical reasons rather than scientific ones. Just as ancient scientists sought order, symmetry, and beauty in their description of the natural world, twentieth-century scientists have upheld local causality. A dichotomy is evident between what the quantum theorist wants to believe (Bell's experiment to show nonlocal influence) and the philosophically sound basis for only local influences; explaining it away as an illustration of quantum weirdness has not generally persuaded scientists to abandon any of the conflicting bases for the dichotomy.[23]

Is there a solution to this dilemma? It seems possible that objectivity for the microworld could be accepted without taking a step backward into classical physics.[24] This possibility is based on an understanding of quantum theory as incomplete (hence, its weirdness) and the hope that scientific knowledge may someday be sufficient to restore determinism. The need for statistical methods would then be recognized as merely a manifestation of incomplete scientific understanding; the ignorance basic to randomness would be replaced by order.

Can Understanding Come from "Weirdness"?

At present, the "quantum weirdness" distinguishing the unexplainable microworld from the rational macroworld is an expression of faith in what quantum theorists believe to be truth. The indeterminacy born of quantum mechanics is sometimes interpreted by physicists to imply that the universe is meaningless with all events being random and due to chance.[25] This interpretation has made human life seem meaningless to some people, with biological processes no longer controlled either by God or by some grand design. The randomness of an indeterminate process has become the basis for explaining how evolution produces changes in living matter. That explanation entails chance errors or "mistakes" appearing spontaneously in the genetic code, so that a physicist might come to the conclusion that life is simply a "disease" of matter.[26]

"Most people find evolution implausible"[27] because errors occurring randomly, "by chance," cannot intuitively be conceived as being responsible for development of life that shows increasing and progressive complexity. Some physicists consider such an idea of evolution too implausible to have been *imagined*; it had to be *discovered*, making it real.[28] In drawing that conclusion they have used the philosophy generated from theoretical physics to comment on the theory of biological sciences. Some physicists' ideas on evolution seem as wild as their concepts of quantum theory, or the "big bang" creation of the universe.

Not all physicists have translated the indeterminacy of quantum theory so literally into a biological indeterminacy. For some, reason, common sense, and intuition prevent acceptance of an extreme version of indeterminacy; for others, some form of a determinate process offers the only reasonable explanation. Thus some scientists must believe, not in chance, but in some unimaginable "miraculous process" which operated to begin an evolutionary process for producing human beings from a series of chemicals out of the primordial soup.[29] Although such an explanation has metaphysical overtones, is it less plausible than viewing the process of increasing biological complexity as a disease of matter?

Atomic Structure; What's New?

Considerable effort has been made in recent years to describe subatomic structure going several orders beyond the atomic theories of early Greek scientists. Some of the most recent information is theoretical and remains to be verified. The early Greek scientists had no means for making observations to verify their theories, but some of their intuitions were remarkably correct, as we have said.

Finding Only Footprints in the Sand

In this century instruments have been created not so much to directly observe the fundamental particles of matter as to detect their properties by examining what they do under different conditions. Descriptions of atomic properties first reduced atomic substructure into nucleii and electrons. Further experimental probing found neutrons and protons by smashing atomic nucleii and examining the particles formed. Dissection of these subatomic particles continued; new phenomena suggested that neutrons and protons are composed of subunits called *hadrons.*[30]

Even hadrons were not accepted as the smallest fundamental units of matter, however. The mathematical imagination of theoretical physicists created even smaller particles, subunits of hadrons called *quarks.*[31] Physicists have never observed quarks and today few believe they will ever be seen; quarks seem to exist only when bound together to form hadrons. The quark arose from a mathematical fiction which somehow worked. Quarks were not conceived from experimental evidence and attempts to split hadrons have not produced smaller particles, merely more hadrons.

There's More to the Mystery

Matter contains a second elementary particle, the *lepton.* The electron is an example of the lepton, differing from the hadron in having relatively weak forces for interacting with other particles.[32] This property allows leptons to exist in a free state, something hadrons with strong interactive forces cannot do.

The third and last elementary substance is the *gluon,* which holds particles together. Each gluon produces one of four interactions between particles. Gluons differ in strength and in the nature of their field of interaction, with such fields being described as having gravitational, weak, electromagnetic, or strong interactions. The strengths of the four interactions are theorized to become equal at very high energies. Such conditions of very high energy, it is suggested, were present at the onset of the universe's creation. "In the beginning" according to the big-bang theory, there was perfect symmetry, including that for the gluons.[33] The loss of that perfect symmetry to produce four distinct interactions in different gluons can be thought of as an evolutionary process.

In spite of numerous possibilities for the ways the high-energy gluons could have changed, it is presumed that what evolved were only the four we have today. All matter and life would be impossible without the four interactions, and the nature of the entire known universe is also completely determined by the precise strengths that gluons have. If the entire natural world evolved because of random events and chance interactions, why should only these four interactions have evolved and why should they have evolved with the precise strengths they possess? The mathematical probability that only four gluons should occur, with their precise forces, is infinitesimally small. Even for those who hold mathematics to be the bastion of *a priori* knowledge, the universe must lose its indeterminacy and randomness.

Perfect Symmetry Makes for Nothing

Determinism gives the universe a reason for the evolution of only the four interactions out of a symmetry with a single force at the onset of creation. That initial symmetry represents a state of complete randomness, absolute disorder, chaos, and maximum entropy.[34] Evolution requires the breaking of the initial symmetry and the appearance of imperfections in the symmetry resulting from loss of randomness; in all of evolution the most complete and complex symmetry-breaking is manifest in the development of successive orders of life.[35] The ultimate imperfection of symmetry is in the greatest complexity achieved, that found in human life. The drive for increasing complexity makes evolution's direction progressively away from perfect symmetry, something that cannot develop from random events. Perfect symmetry requires all interactions and events to be random and the product of chance, an indeterminate situation. The development of increasing complexity in the evolution of living matter indicates that the world is deterministic.

Gravity, the Greatest Mystery of All

Of the four forces or interactions, gravity was the first to be investigated, beginning in the seventeenth century; as a phenomenon it is understood well although the physical basis for gravity is completely unknown. Gravity is suggested to be manifested as waves, but they have never been detected.[36] Despite the fact that no exchange particle or quantum has been identified as mediating gravitational attraction between bodies of matter, scientists have given the name *graviton* to the hypothesized quantum. The graviton provides the glue that is the attracting force in a gravitational field.

It may be that gravitons are too weak for detection by present-day technology. If gravitational waves are ever detected, gravitons are expected to travel at the speed of light; the dogma of scientists includes the belief that nothing can travel faster than that. At any rate, methods for their detection are designed on that premise. If gravitons do move in waves, their speed might turn out to be far greater than the speed of light, however. But scientists have theoretical and philosophical reasons to believe that it is not possible to observe anything at such speeds, hence, such a graviton would not exist.

The universe might behave more ideally if gravitational forces were able to cause more nearly instantaneous effects. The speed of light is relatively slow considering the scope of the universe. According to quantum theory, events in the near vicinity, local events, could instantaneously affect distant happenings at the opposite end of our galaxy. Gravitational interactions throughout the universe may be such nonlocal instantaneous interactions, which may preclude an easy detection. The fact that nothing has ever been detected to move at speeds exceeding the speed of light does not preclude that such speed might exist.

To the modern astronomer, the overall appearance of the entire known universe shows a remarkable homogeneity, so that the appearance of all parts distant and near is uniform.[37] The basis for such order and symmetry is completely unknown, although scientists seek a cause within physical laws. In order to maintain cosmic homogeneity, overall symmetry must be maintained by interactions of all components of the cosmos. Thus, the composition and interactions of one part of the universe would influence those of distant areas millions of light years away. Such nonlocal instantaneous interactions are suggested by quantum theorists but denied by the common sense and reason of classical physics. Interactions of this type might be explained by either (a) unknown physical forces that travel faster than the speed of light or (b) forces that can never be understood scientifically, relegating them to the metaphysical.

Scientists believe gravitons exist and that they will be found to be similar to the other three forces. That may make it possible to develop a unified theory for describing the graviton's origin and evolution from the onset of creation.[38] To believe that gravitational force might be explained by anything other than gravitons would force scientists to give gravity a metaphysical basis (like some kind of telepathy), something positivistic scientists are not prepared to do. Positivism maintains that the only valid kind of knowledge is scientific and that such knowledge consists in the description of the invariant patterns in space and time of observable phenomena. Positivism rejects any other source of knowledge.

Unifying Using New Mysteries

With a mathematical description of the essential units of matter (quarks, leptons, and gluons) in hand, physicists have tried to explain their interactions mathematically; the result is called *quantum field* theory. Quantum field theory made no sense until a mathematical sleight of hand was performed which seemed to work.[39] This mathematical cheating, called the "renormalization procedure," is comparable to manipulations that would place an infinite weight on a scale and then reset the scale by an infinite amount to get a finite weight. Since the infinite addition and adjustment cannot be calculated, the adjustment to a finite weight becomes completely arbitrary.

Rescuing quantum field theory through a renormalization procedure may represent progress in understanding the universe. If so, it has come only by exploring new mathematical concepts, not by directly trying to explain experimental results. Yet physics has always been concerned with discovering the truth about physical entities, using mathematics as a means to that end; in quantum field theory it has become an investigation of a purely mathematical structure for the world.

A physicist might say that God has used beautiful mathematics in creating the world.[40] Perhaps it is not surprising that a kind of ultimate understanding of the universe would be sought through mathematics. People have been enthralled with symmetry for centuries, believing that geometrical and mathematical symmetries are "beautiful" and based on beautiful sets of ideas.

With the idea that the natural world is based on such beautiful mathematics, scientists developed the *gauge field* concept to describe a symmetrical ordering for basic particles in space. This concept was forced on nature to explain the interactions of particles in a field. It didn't work, however, until another mathematical renormalization procedure was applied.[41] Having done that, some physicists saw nature as indeed based on the beautiful mathematics of the gauge field symmetry. The gauge field theory may have given physicists a deep clue about the structure of material reality.

The quantum theorist demonstrates magnificent confidence to trust in the unseeable, unknowable, and unprobable quark; to believe that the design of nature is structured by mathematics, a human invention inspired and developed to deal with abstract ideas; and to accept the observer-revealed reality offered by instruments of human design as the true reality of the natural world. The view of material reality coming from such a position can be built only on faith.

Mathematics for Believing the Unseen

Scientists maintain that there are causes or explanations for all phenomena in the natural world, and that they can be revealed with rigorous application of reason and knowledge as emphasized by positivists. Metaphysical explanations are not allowed; yet that seems inconsistent with intuition and imagination playing such an important role in developing new scientific concepts and applications.[42] Positivists are uneasy with intuition and imagination because they skirt the edge of metaphysics. Yet the imagination of modern physicists produces ideas with an element of craziness, well-grounded outrageousness, or weirdness—features not expected from a positivistic approach.

Modern physics using a powerful imagination constructs theoretical representations that become works of fiction. Although physics has lost its causality and become indeterminate, it is still believed to be a more reliable source of truth than any narrative knowledge because it uses (or is used by) mathematics. But what if the denial of determinism, allowing for only a statistical description of the natural world, should lead to the end of a mathematical description of nature, traditional foundation of the scientist's faith?[43] So far, faith in mathematics perseveres and contributes importantly to human fictions. The birth of the universe, for example, is a story built and sustained by mathematics.

The Ultimate Science Fiction Story

The universe is said to have arisen out of a vacuum as a "re-expression of sheer nothingness." Even in its present form it could be said to be equivalent to nothing: if all the energy in the universe were added up it would amount to zero.[44] Such beliefs, though supported by mathematics, are totally inconsistent with reason and common sense. The biblical creation story describes a similar condition before the universe's formation but it was not intended to be a scientific account. The account begins a "salvation history," giving people a concept of God's power expressed in the act of creation. The biblical creation account, rich in intuition and imagination, seems strikingly similar to the physicist's account, which, even though explained by mathematics, also has a metaphysical basis.

People gain fulfillment through faith and feelings. For many, fulfilment is found in mathematics, beautiful in its order, symmetry, simplicity, and coherence. Does mathematics have a fictional basis? Is it a human invention inspired by an innate capacity to deal precisely with abstract ideas? Its basis would then not differ substantially from those of art, literature, and music.

The human invention of a beautiful mathematics has been applied to the natural world and made to fit, at least well enough to support a belief that mathematics is the *a priori* bastion of knowledge. Yet the fit between our mathematical theories and the complex, messy, real world is never perfect. The urge to make them fit is a desire to give the natural world the beauty, symmetry, and order of our mathematics. That is, we want the universe to have physical laws that can be expressed mathematically and are invariant, simple, universal, and complete in their explanation of phenomena; we also want those laws to be demonstrable by observations or experiments.

Physical laws give the universe qualities that humans believe it should have. Such are inconsistent with a creation that could be characterized as being random, spontaneous, and without purpose. On the other hand quantum theory and modern theories of evolution do seem to convince some people that the universe is random, indeterminate, and purposeless. That kind of creation is indifferent to human needs; the more one delves into it scientifically, the more pointless and meaningless it appears to be.

Call the World's Determinism a Cosmic Code

That our universe is *not* one of disorder and purposelessness is recognized by cosmologists who express belief in a cosmic code—a message in mathematical code describing the universe.[45] Belief in such a code is not really new; it was expressed by early Greeks and later by scientists such as Francis Bacon. Plato thought everything in this world was an imperfect representation of a perfect "design."[46] Many scientists consider the major challenge to civilization to be the mastery of the discovered contents of the cosmic code.[45]

Physicists' concepts are metaphysical when not based on scientific experience. A cosmic code is in the realm of the metaphysical, not the realm of human experience. With a cosmic code the universe becomes the material revelation of its message, one that is now programming all human social and economic development as it has from the time of the universe's creation. The unseen in the metaphysical code is influencing the seen of the material world. The unseen may be remote and cause change instantaneously. With that understanding, the universe has regained its determinism, objectivity, purpose, symmetry, and causality. The "grand design" is returned from the mathematical realm to that of the mystic, metaphysical, or theological. It is by intuition and imagination that we are brought to this conclusion—after scientific knowledge and reason fail to reveal purpose or meaning for the natural world.

Symmetry to Asymmetry, a Progressive Revelation

The purpose of the world is unfolded in its evolutionary development. In all matter, asymmetry evolves from the symmetry present at the creation, whether one believes in the Bible, the big bang scientific theory, or both. The most important idea from the golden age of reason may be that scientific knowledge shows an assymetry developing from symmetry, order from chaos, and complexity from simplicity. That evolutionary process is inconsistent with indeterminism but consistent with determinism pointing to a design or to God.

Chapter 7

The Fabric of Theory and Myth

"Theories have always been expressive of the myth or metaphysics of society, and have therefore been part of the internal communication of that society. Society interprets itself to itself partly by means of its view of nature."[1] But reaching understanding cannot be sought only in society's views developed from science and technology; it requires a cultural tradition that ranges across the whole spectrum.[2] What has been the contribution of science and technology? History shows continuing progress in developing better technology for improving life's quality. Is there also such a cumulative or progressive nature to scientific theory?

New Idols Replace the Old

Many scientific theories of the past have been superseded. Our present theories on cosmology, evolution, and quantum physics may also be tenuous, even though they have some empirical support. If scientific theory were based on empirical certainties which did not leave openings for innumerable explanations, it could be said to be moving toward truth.

Most scientists believe there is vast evidence to support such modern theories as the big bang theory of the universe's origin and Darwinian evolution. But in fact, philosophers of science see an underdetermination of both theories by empirical information; that simply means that other theories can be found to fit the observed facts. According to philosopher Mary Hesse:

> Since the truth of a theory can never be guaranteed, there is always a multiplicity of theories that will fit the facts more or less well, whose credentials will rise and fall with culturally accepted norms as well as with experimental developments, and which will each in all probability eventually be refuted and rejected from serious scientific consideration.[3]

This is not a new idea; Friedrich Nietzsche argued that it is "possible to conceive a reality that can be resolved into a plurality of fictions relative to multiple standpoints."[3] The decision to select one theory as fitting the facts better than any other is based on extra-empirical criteria. Those ever-present external factors behave as a social conditioning of the theoretical belief systems of science to determine what will be accepted as knowledge in any culture.

Social determinants influence theory selection in any culture. The structure of human society includes cultural manifestations in ideas, beliefs, religions, art, and knowledge. The determining criteria of theories in natural sciences come from theological, metaphysical, and ideological sources.[4] Scientists as well as philosophers have always sought theories showing beauty, simplicity, symmetry, and balance. Those metaphysical foundations remain acceptable for theory construction, but overtly theological grounds, once important, are no longer acceptable criteria determining theory choice.

Cultural conditioning manifests in all human endeavors

the influence of the unconscious on the conscious, the preconceptual and nonconceptual in the conceptual, and the irrational at the very core of the rational.[5] The knowledge sought in these endeavors is not objective but is intimately derived from human interests and is the basis for people's narrative (as opposed to scientific) knowledge to include that of science's presuppositions; although paradoxically it is impossible to judge any validity for such presuppostions on the basis of scientific knowledge.[6] "Science is incapable of legitimizing itself, as speculation assumed it could."[7]

Individuals are nurtured and develop in a society before any one of them becomes a scientist. The cultural conditioning of that society determines how theory choice will be made, long before a particular individual begins accumulating empirical data and making decisions for theory selection. Thus, scientific theories are made not only by empirical data from observations and by the logic of rational argument.

A theory may be drastically modified or even discarded as fake if sufficient adjustments are made in the extra-empirical criteria or determinants of good theory.[8] The "goodness of fit" between empirical data and logical reasoning may be unable to overcome the influence of social conditioning. People sometimes reject empirically successful theories lacking beauty, symmetry, and order in favor of theories built on little or no empirical evidence but with those desired metaphysical traits. Scientists may not accept the idea but ". . . in the sciences of man, a sense of 'truth' that is not merely pragmatic may be derivable from prior commitment to value and goals."[9]

Changes in cosmological theory by extra-empirical determinants may be more likely than changes resulting from new knowledge. The possibility of uncovering new information on events during the dawn of the universe is very slight. Theories of creation are more likely to change because of new ideas in what people find pleasing to the eye of the mind, such as beauty, balance, or creation out of nothing.

Comfort in "Sound" Theories

Progress for scientific theory is regarded as advancement toward the ultimate truth of the nature of life and the universe. The hitch is that to identify such progress requires some absolute criteria to know when truth is gained. That is not possible when the generation of empirical data can be explained by not a few but many different theories. Also considerable "mythmaking" enters into the construction of scientific theory because descriptions of the world known to be false are used to construct such theory.[10] Ascertaining scientific truth is also difficult when concepts are in part determined culturally through extra-empirical criteria. The empirical evidence for a big bang theoretical creation of the universe is as meager as for any scientific theory; yet it is popularly received. Big-bang theory may be pleasing because it creates the universe out of nothing, in keeping with the biblical account of creation but not with the long-time concept of matter not being created or destroyed. Cultural beliefs do dictate everyone's view of all creation.

Thus, there is no evidence of progress for scientific *theory* in answering the truths humans seek; that should not be expected of science. The idea that our universe can be described by universal law-like generalizations is actually based on a belief or intuition. People *believe* in the existence of an economic, coherent system of laws which show symmetry, simplicity, analogy, and conformity with certain *a priori* conditions or metaphysical postulates.[11] Ever since that belief developed from metaphysical considerations through the logic of deduction, it has been used to force empirical findings to fit the mold of our perceptions for the universe. Scientific theory has acquired such prestige that to many it is the only reliable, universalizable mode of knowledge. Yet not everyone accepts the unique rationality of modern science as *the* road to knowledge.[12]

With a "Tainted" Objectivity

Scientists would like to believe that scientific theory is always objective, universal, and showing a good fit with empirical facts. That would make it different from theory in other areas of inquiry, such as the social, religious, or political. Scientific theory in the basic sciences may be more guilty of using false descriptions of the world in its "mythmaking" than sciences admittedly less objective in their methods, such as the sciences of human behavior.

To some philosophers, methods used in the basic sciences for theory construction are far from objective and are little different from those in the "softer" sciences. Paul Feyerabend believes that scientific theories and arguments of even the most basic sciences are closely analagous to the circular reinforcement of beliefs, doctrines, documents, and conditional experience found in some religious groups and in political party lines, with their associated techniques of propaganda.[13] People use such circular arguments to describe the world in a way they think it should be; we are biased because we belong to the world we wish to interpret. That should not be surprising since people learn a cultural basis for theory construction long before having to develop a scientific theory. By then it becomes difficult to discriminate and eliminate inappropriate extra-empirical criteria for theory formulation and selection.

Observations are "tainted" by the time they are recorded. Far from being empirical and descriptive of the world, even *observations* are biased by human choices of what to observe, how it should be viewed, and what one expects to

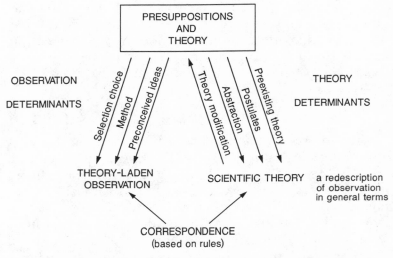

Figure 3. Schematic showing circular interpretation, reinterpretation, and self-correction of data, with data in terms of theory, theory in terms of data.

see (Figure 3); the observed is preinterpreted from the beginning.[14] Since those choices are determined by currently-held scientific theory, the observations are already theory-laden. We attempt to understand more on the basis of what we already comprehend. In addition, the theories formulated are based on (a) preexisting theories that are no longer acceptable, possibly for some nonscientific reason; (b) postulates with little empirical support; and (c) abstraction, arrived at by metaphysical means of intuition and imagination. Thus, theory determines the selection of observation determinants, and the resulting theory-laden observations dictate redescriptions as new theories—a circular pattern of interpretation and reinterpretation. Observation statements carrying theoretical implications or based on the substance of a theory are unreliable evidence to support the theory in question.

Science, Learning, and Survival

The goal of science is to assist people in learning the best ways to survive and prosper in their environment.[15] Some basic conditions are essential for that learning to be successful. First, there must be a possibility of detailed testing of theoretical and technological achievements so they can be reinforced by observations.[16] Although that possibility exists people often ignore the outcome of such testing. For improving technology, detailed testing is the norm. For achieving cultural progress, testing may not be done or its results may be ignored; history frequently has to repeat itself. Technological progress rests on ample empirical evidence and detailed testing. What remains to be done is to decide on what is most needed for survival.

Second, people's environment must remain sufficiently stable to reinforce learning by detailed testing.[16] The environments of large urban areas are constantly changing due to the pressures of increasing population; stability for adequate testing of social theories is seldom possible. Unstable environments are also created by the rapid denuding of forests in large areas of South America, Africa, and Asia; changes in climatic conditions essential to life may be occurring too fast for people to cope.

The third condition for learning by observation is recognition that observed phenomena may be artifacts of the observing process which do not represent the real situation.[17] Both subjective factors of the observer and methods for observation can determine what is perceived. An observer-created reality is little help in learning how to best survive in our environment.

We tend to think that we have learned a lot. Even in such nonpragmatic areas of interest as cosmology and evolution, these three conditions of successful learning and control have been imperfectly satisfied, however.

Seeking Sources of Truth

Scientific theory is expected to be coherent with sources of "truth" in the logical and pragmatic realms.[18] Other coherence conditions, not universally recognized as sources of truth, include the economy, simplicity, symmetry, order, and beauty favored by human intuition. Other conditions described as being innate add a metaphysical dimension to the approach to truth through development of theory.

The most discredited coherence conditions are those that make some "value" a prerequisite for accepting a scientific theory. Examples might include the earlier theological objections to a sun-centered universe and the moral objections to the Darwinian theory of natural selection. Those theories were unacceptable to nonscientists because of adverse affects they were thought to have on the general population.

For the *scientist*, a theory can have no value considerations if it is to be objective and therefore approach truth. In science the goal is to be completely detached from nature in order to study its manifestations and comment about it. In this way conclusions can be drawn that are objective—that is to say universal, the same to all people. Without detachment, scientific conclusions would be influenced by personal values, losing their universal quality.

A scientific theory's overall validity is determined by the truth-value of coherence conditions as well as by its correspondence with facts (the validity of observations supporting it). Theories in the physical sciences seem closer to truth because they appear to be free of value-determined coherence conditions. Yet all the properties of the most basic particles of the universe *have* value since they must be precisely as they are in order for any life to have developed,[19] something we all must value. With that "choice" of those properties coming from an infinite number of possibilities, value-determined coherence conditions seem logical in explaining the physics behind human existence. The alternative explanation for the highly defined properties of those particles is that of chance appearance during the creation, an improbable and unacceptable condition for the scientific truth of basic particle properties.

If current scientific theories must be as subject to radical conceptual change as past theories have been, how can there be an accumulation and convergence of scientific theories towards truth? Science does show progress in being able to correctly predict observable phenomena, however, in spite of what must be at best a poor correlation between scientific theory and truth. Perhaps our success in scientific technology and prediction, without a firm basis of scientific theory, could be called a miracle. According to Mary Hesse:

> Putnam has argued that without the assumption that there is some accumulation, progress, or convergence of scientific theories towards the truth, the success of science in correctly predicting observable phenomena would be a miracle.[20]

But she goes on to show that successive theories are in radical contradiction of one another, and whereas:

> Facts of lawlike structure and similarities of nature between physical systems seem to be maintained and to be cumulative. Theoretical interpretations of what the natures of these systems are, are not.

In essence Hesse is saying: although science makes technological progress it makes no theoretical progress. In this way scientific knowledge differs little from traditional narrative that makes no pretension of cumulative

progression.[21] Most people believe that scientific knowledge generates technology, a tool for people to cope with
reality. Modern science does not enable this coping because it corresponds to truth, "it just plain enables us to
cope."[22]

Scientists may admit to being not as successful as they
would like to be in any of the sciences:

> For it is not a convergence of ontologies approximating
> better and better to a description of the true essence of the
> world, to a final delineation of the pre-Socratic ideal of
> "what the world is made of." It is rather an instrumental
> control, as pragmatic as our desire to have true, controlla-
> ble predictions. Science is after all the Martha, not the
> Mary, of knowledge.[23]

The pragmatic predictive evaluation of observable
properties of atoms and molecules can make progress
even at a time when theoretical explanations by Newto-
nian, Daltonian, quantum, and relativistic field theories
provide very different descriptions of the basic structure
of matter. Successive theories of the basic structure of all
matter show no convergence but oscillate between continuity and discontinuity, field conceptions and particle
conceptions, and among different topologies of space.[24]
Although we know that water is composed of discrete
molecules of hydrogen and oxygen in definite proportions, ". . . we are not able to specify in ultimate terms what
exactly molecules and atoms of water, hydrogen, and
oxygen *are* . . . ."[25]

Natural science has dominated the realm of reason,
knowledge, and truth for over three centuries and has
rapidly expanded technology for controlling and predicting in this century. Since its success, technology is accused
of enslaving people to the modes of social organization it
generates.[26] Has science led us toward truth?

Seeking Isolation to Gain Objectivity

Truth must be objective and universal to be recognized and accepted by all. Value considerations and subjective interpretations are considered subversive influences on the objectivity that scientists maintain is essential for science to have truth. To view our existence as a purposeless life in a meaningless universe is to concede a value consideration and make an interpretation; to view that conclusion as "objective" must also be meaningless since everything must have no meaning in such a universe. Consequently, it makes no sense to believe that humanity is estranged from nature in an indifferent universe. Such isolation is a myth based on no empirical evidence; it represents society interpreting itself to itself at least partly by means of its view of nature. People cannot separate themselves from nature and examine it disinterestedly, so there can be no "objective" theories on the natural world.

"Infected" Theories

The integral relationship of humanity and society to nature as a whole will continue to "infect" scientific theories. Interpretation of ourselves to ourselves arises most importantly from creative imagination, not necessarily from experience of the world. Truth claims for any society are determined by theoretical dogma fundamentally unproven by any empirical evidence.[27]

In the modern Western world, the concept that humanity exists in isolation and meaninglessness developed out of physical ideas on fundamental particles and fields, on a space-time continuum, and on forces of interaction. Our concept of humanity also includes the theoretical dogma of evolution, ecology, and genetics. It is not always possible to separate knowledge relating to technical control from

that expressing how we understand ourselves. Categories
of scientific theories such as functionality, selection, and
survival are influenced by humanity's view of itself.

In contrast, the truth claims of many other societies are
rooted in theoretical dogma based on spirits, witches,
telepathic communications, etc.—other expressions of
how societies see or interpret themselves. Westerners
think their theoretical dogma has a firmer scientific basis
than that of primitive societies.

Modern society obviously places a value on exploitable
technical control. The view of itself that enables society to
exploit technology does not necessarily "square" with objective and thus value-free science. Society is not detached
or disinterested in its own survival. Objectifying is self-
defeating and is linked to the interest in extending instrumental control over the individual and society.[28]

Is My Theory Better than Yours?

A scientific theory has no claim to objectivity if other
theories can explain a given body of facts and if certain
value-considerations have led to its selection over the
others. Those considerations might be based on the
theory's ability to predict phenomena in the universe, on
pragmatic interests in the technology that issues from it, or
on a close agreement between it and the way we believe or
want the universe to be. Another value judgment determines which of these value considerations should be most
important in theory selection. With any conflicting concepts there is an incompatibility in the way the beliefs of
each becomes structured since each has internal to it its
own standards of truth and justification.[29] There is no
independent standard of rational justification to justify the
choice of one set of beliefs rather than the other. Facts
alone cannot establish truth-value for a theory.

People Do Not Live by Science Alone

Success in predicting and controlling the natural world led scientists to study humanity as a purely natural phenomenon to be described in terms of behavior, language, laws and customs, and approaches to economic production.[30] Attempts to adopt the pragmatic criteria of predictability and technological control have not been successful for developing general theories in the social sciences, however. If humanity is viewed as a natural product of evolutionary forces based on random variation and natural selection, human destiny means no more than fulfilling a role where the fittest survive.

If science does not provide a satisfying basis for meeting human social needs, where can such a basis be found? Could objective truth leading to social fulfillment be found in common commitment to innate, transcendental, or religious values?[31] With consensus on that commitment, truth is accepted as the will of all, or of the majority.[32] People committed to metaphysical values can hardly expect science to add to those values for understanding, predicting, and controlling human actions. On the other hand, people with no commitments to metaphysical values have hardly found science to be of much help in the social domain.

Is Truth What Is Agreed to Be True?

By consensus, truth being what people agree to be true, dogma could become objective and universal. People select and discard beliefs according to sectional or local interests and cultures. Establishing truth by ideal consensus is recognized as a method for both empirical and interpretative science.[32] Truth in both is gained by interpersonal communication under conditions of ideal freedom and justice for people with a commitment to discourse. With consensus, social interactions are based on temporary contracts with those supplanting permanent institutions in the professional, emotional, sexual, cultural, family, political, and international domains; the temporary contract is favored because of its greater flexibility, lower cost, and creative turmoil.[33] Does that provide a flexibility for freedom from commitment, a cost of minimal commitment, and a disorder to drive an evolutionary process?

Ideal consensus is considered a supreme achievement of evolution and a transcendental (beyond understanding) condition of commitment to discourse.[34] But how could ideal consensus be a product of evolution and give purpose to it? Even more difficult, how could the cooperation and understanding of people committed to consensus develop by an evolutionary process based on random variation and natural selection? How could evolution have developed a transcendental condition necessary for consensus? Consensus is inconsistent with an evolution based on survival of the fittest; neither is consensus as an ethical argument one that could develop by Darwinian evolutionary processes.

Neither scientific theories of the physical or biological worlds nor consensus or other human schemes has developed a system of ethics to encompass all humanity; people cannot know or prove what is true by any system developed by the human mind. An ever-increasing trust in strictly human systems has persisted despite boom and bust, famine and glut, wars and holocaust, Hitler and Stalin, Hiroshima and Vietnam, pollution, drug culture, terrorism, and the shadow of nuclear madness that could end human history.

The dubious achievements of our systems suggest that the intellect as a privileged human function might *well* be a product of purposeless evolution based on natural selection and survival of the fittest. Believing itself to be a product of Darwinian evolution, it would not be surprising

The Beginning of Truth

Anthro primo:
"Only when our beliefs came to be
true did we learn to communicate
for in order to communicate most of
our beliefs must be true. Listen and
hear many truths."

Anthro secundo:
"I had no problems with truth until
I learned to listen and talk."

if the human intellect considered it appropriate to use consensus to condone and encourage any of the excesses that have plagued humanity this century. To do what is right in our own eyes translates easily into survival of the fittest.

Where an individual cannot know or prove what is true, consensus of a group of people can never be the way to truth; truth must be accepted as such by all. We begin to learn truth only when we recognize the limits of possibility for the systems we create. Neither success in technological and instrumental control, nor scientific theory, nor consensus opens truth to us. Yes, even "consensus has become an outmoded and suspect value."[35] This may be in part because cultural relativism, the product of consensus, bears a deep irrationalism, a denial of the possibility of thinking.[36]

God: Once the Source of Truth

At one time human beings sought truths from their gods; for many people now religion is hardly a source of truth. Some have discredited religion by emphasizing how it has failed in its attempts to state objective truths. Subversion of religion gained momentum when the religious truth of a human-centered geocentric universe was contradicted by discovery that human beings exist on a planet of no special significance in a heliocentric universe. It continued when the religious truth of God creating humans from nothing was shattered by evolutionists arguing that we descended from animals. Then the religious truth of our creation in the image of God with the possibility of having the spirit of God within us was lost with the Freudian idea that the animal, not the spirit of God, is what is still in us. Science was able to argue that since religion is unable to state objective truths, people should look to science for those truths, and especially those which answer basic human questions.

Myths: Theories with Infinite Improbability

A truth is a reality for every person throughout the world; in contrast, a myth is something very unreal, something infinitely improbable. For the scientist, determination of truth cannot be by subjective criteria or merely by consensus; it must be established by observations showing a very high probability that something is real. Human imagination is the fabric of both theory and myth; only probability distinguishes between the two. Theory cannot be merely what we think something should be; conversely, myth cannot be attributed to whatever we do not find satisfying. Are scientific theories on human origins and development more than myths?

Humanity's Appearance; More than an Accident?

Naturally curious about their origins, people have attributed human existence to one of two different causes. Some explain humanity's appearance as the product of an evolutionary process beginning in a primordial soup and amounting to a matter of probabilities. Others explain human existence with a creation story such as that found in the Holy Bible.

Beyond Theory and Myth, Probability for a Design

Rather than appearing "by chance," human beings may have appeared because of a specific design; we may have been destined to appear. The plans for such an appearance could have been established at the time of creation. Cosmologists describe the events of creation in the big bang theory; theologians call creation an act of God.

As we have noted, it seems unlikely that the very exacting conditions for the physical laws of the universe could have been the product of chance. With minor variation in those laws, no form of life as we know it could exist or have developed by an evolutionary process.[19] Life's very existence is predicated on severe constraints on the physical laws of the universe, which, if unchanging, must have been established at the time of creation, long before the appearance of any living matter. Again we ask, were those constraints the product of a design?

Specific earth-solar relationships are essential to human life.[37] If the earth were five percent closer to the sun, life could not have evolved; if the earth were one percent farther away it would be in a perpetual ice age. The age, star type and brilliance of the sun is also critical in determining the temperatures on earth and in sustaining, through millions of years, conditions necessary for life to evolve.

On a more cosmic scale the entire universe is considered to be homogeneous or uniform in structure with inhomogeneity found only on a smaller scale in galaxies.[19] That overall uniformity is essential for life on earth. If one accepts the big bang as one's picture of creation, the existence of this universe depends on exactly the right velocity of particles enabling them to escape from the gravitational force of other particles.[38] If the first velocity (recessional) were greater than the later escape velocity, no matter could have aggregated into planets, stars, and galaxies. If the recessional velocity were less than the escape velocity, the universe would collapse before uniformity was reached.

A single "arbitrary" value for recessional velocity allowed our present universe to be created. That one condition appeared out of an infinite range of possibilities—and only one value could lead to our world and human life. Was it chance that this one value for recessional velocity appeared? That infinitely small probability offers support for the concept of design.

Again, we have noted the four fundamental forces affecting the particles of all matter: gravity, electromagnetism, the strong force, and the weak force. The properties of these forces must be exactly as they are in order for the development and continuing existence of the universe—and life.[19] Their magnitudes are completely predictable and not subject to probabilities, suggesting conditions that are very determined. The invariability and specific strengths of these forces were absolutely essential in a universe where human life was to appear. Other things could be uncertain in that universe but not the properties of these four forces. The fact of the four forces with their very specific properties is astronomically unlikely to be a product of chance; as a product of design, they form part of the "cosmic code." Design rather than chance is more probable in explaining the properties of the universe.

No one dares to speculate that the appearance of the properties of the four interacting forces and no others can be attributed to chance. In fact, scientists look for a design or pattern in attempts to develop a unifying theory to explain all matter. They seek a theory unifying the interacting forces which would demonstrate unity, interaction, beauty, simplicity, and symmetry. For their search they draw on their imagination, the fabric of theory and myth.

The Myth of Deism Revisited

The search for a design and unifying theories to explain the universe is motivated by ideas similar to ones held by deists in ancient times. Modern deism allows for a design, plan, or even God to give beauty, simplicity, and symmetry to the universe on both the macrocosmic level and the subatomic level. After the universe's creation, however, modern deists believe the creator to be a disinterested observer, allowing the universe to develop and run merely by chance.

Modern scientific theory (or myth) may have placed human beings in an indifferent, uncaring, and meaningless universe but it may be that knowledge of the absolute and invariable properties of basic particles and interacting forces will put humanity back into a meaningful world. The scientist's zealous conviction that nature possesses order, beauty, and simplicity may not require a completely deterministic universe, but at least a universe designed and running according to a plan.

Chapter 8

Science: Promise of the Covenant

Moral virtue and wisdom enable human beings to enlarge the limits of their freedom. At a very basic level, gaining freedom from passions and desires leads us to readily accept doing what we ought to do and not doing what we ought not to do. To seek only *needs*, subjecting personal interests to the needs of society, is the essence of moral behavior.[1]

Science and Technology Increase Human Desires

Seeking only natural needs, people strive to gain their natural rights. Very few scientific achievements provide us with our natural needs. Instead, the accomplishments of science and technology give a small fraction of humanity much of what they *desire* regardless of whether those things are really good in this world. Technology's success has been counterproductive by greatly expanding the range of people's desires; satisfying those desires has led to neglecting the real needs of millions of people.

Science currently does not endeavor to satisfy everyone's basic needs. Natural needs are the basis of our natural human rights, what we need in order to discharge our moral obligation to seek everything that is really good for human life.[1] If the natural needs of people universally were satisfied, that would indeed be a remarkable democratizing of human civilization. By the seventeenth century science had become one of "the greatest democratizing forces in the history of civilization" by enabling many people rather than a chosen few to discover the "truths" of science.[2] But since then science has never become a great democratizing force in a broader way by guaranteeing the benefits of science for all humanity; human rights are not more universally held now than at other times during the history of civilization.

Any view of current science as a democratizing force is value-laden; it is neither objective nor universal, making it a theory or opinion rather than a scientific truth. Indeed, an alternative, admittedly value-laden theory, no less underdetermined, can be offered to explain where democratizing forces should be sought. Science *could* be a great democratizing force, but its promise was not meant for that. Religion and philosophy, which cannot develop from scientific theory, can be the real democratizing forces. Science's promise is for human survival.

Human Beings Need Science to Survive

Science is essential for human survival, and "if modern science, technology and medicine were to vanish from the world or to be severely limited, many hundreds of millions would die . . . . Science has made itself indispensable for any future world society, no matter how it may evolve or be organized."[3] Science is not indispensable because of its theory; its prediction and instrumental control are what we need.

Science's Perception of Its Accomplishments

Most people do believe that science is of immense value. Its greatest value could be to provide humanity with its natural needs, representing our most important natural right. The priorities of science were recently evaluated for nonscientists in a publication of the American Association for the Advancement of Science.[4] In that publication, twenty of the most important scientific discoveries of the twentieth century were selected by a consensus of leading scientists. The publication emphasizes the belief that "the rise of science may be the true hallmark of our times, the most potent social force at work in the world today."[5]

The limitations of science as a new religion and a new philosophy are evident in what were chosen as its greatest achievements in this century. Four of the twenty scientific discoveries of science were related to chemistry: the development of polymers or plastics, drugs to control fertility, drugs to modify central nervous system activity (tranquilizers, for example) and the awareness of drugs and chemicals as important toxins affecting biological function. Four other discoveries were related to electronics, enabling people to communicate instantly by way of radio and television and to analyze and process data with computers, aided by sophisticated statistical tools to interpret data.

Other discoveries included technical developments in recognizing and managing important medical problems, the science and engineering of aviation, the mixed blessing of nuclear fission, laser technology, genetic technology to increase world food production, tests to evaluate human mental ability and achievement, and unraveling the structure of the genetic apparatus.

Three other discoveries were not technological achievements but are recognized as progress in scientific theory. They include the theories of Einstein formulated in 1905 and still being examined for validity by scientists today. Another theory, based on the discovery in Africa of a skull from a young primate argued to be an early ancestor of humans, links human beings to nonhuman origins, crystallizing the evolutionary concept. Finally, the severely underdetermined big bang theory is said to explain the universe's creation.

Satisfying Wants and Desires, Not Needs

If moral behavior is the real good humanity is to achieve, one might question how such discoveries have helped human beings to realize their natural needs universally. Most of the developments in the list provide prediction and instrumental control for gaining human desires whether or not they represent a real good. The discoveries in the list that can contribute to people universally by fulfilling natural needs are the genetic technology to increase world food production, the broad-based knowledge needed to control human population growth, and the biomedical technology to prevent decimation by constantly emerging forms of new infectious diseases.

The scientific theories in the list are said to have shaped society in the twentieth-century. The new theories have indeed shaped our lives by becoming the new source for *a priori* knowledge and truth, thereby replacing philosophy and religion by science. New concepts produced an indeterminism making it difficult to use a design or God for explaining human life with any meaning. If science cannot give life meaning how can there be value in meeting human needs universally?

Science's Perception of Its Promise

Scientists selected by the American Association for the Advancement of Science also chose twenty-five discoveries that could change our lives in the future.[6] They included future discoveries in biomedical technology to answer questions about biological development; create body parts to replace diseased tissues; create cancer experimentally to learn the whole chain of events in malignancies; and create new forms of therapy for treating genetic defects and for using the body's own proteins, manufactured by processes in genetic engineering.

Other desired technology included computers capable of more sophisticated problem-solving through innovations in software; a variety of new plastics and ceramics to replace today's conventional materials of metal and wood; new catalysts to open up the possibility for new chemicals of all kinds; and technology to improve communication and travel. Much hoped-for were discoveries explaining the brain and behavior on a biochemical basis, to aid in developing drugs for managing a variety of neurological diseases.

Other hoped-for discoveries would add to our understanding of current scientific theories established by (a) Einstein, (b) scientific cosmologists on the beginning of the universe, (c) high-energy physicists on the fundamental particles of matter, (d) evolutionists on the time patterns in evolutionary process and on the genetic differences distinguishing one species from another, and (e) mathematicians. Future discoveries in genetic engineering are also needed to increase world food supply.

Hardly anyone would question the value of such discoveries for the benefit of humanity. There is some question about the nature of their value, however. Will they be universal in meeting the natural needs of all people, or will their primary effect be to satisfy the wants and desires of a few people? In the past, science has created new desires which a minority could fulfill at the expense of keeping humanity in general from realizing their natural needs. What we really need is scientific technology that optimizes the opportunity for all humans to live a healthy life and to obtain essential nutrients. None of the other hoped-for discoveries should be given a priority that would jeopardize the realization of these basic natural needs for all human beings.

Such priorities stem from a sound moral philosophy.

For moral philosophy to be sound it must be based on the facts of human nature. Yet none of those facts have a basis in scientific theories about cosmology, evolution, or the nature of fundamental particles.

Truth Must Be Universal; Does Science-Technology Seek Truth?

The truth-claims of science are based on its objectivity and universality. Can science claim truth unless its immediate goal is to satisfy the needs of all human beings?

The efforts of technologically advanced nations are not designed to meet the natural needs of humanity. In the past, those needs could be disregarded with little fear of repercussions, but that has changed. Efforts to satisfy the desires of a minority not only deny the majority of humanity what they need; those efforts also ignore the potential of jeopardizing the natural needs of *all* people, including the minority presently occupied with their own desires.

The demand for scientists to satisfy *wants* rather than *needs* has become unrealistic. Many people believe that medicine will eventually be able to replace human body parts like replacing defective parts to repair a machine. To some extent such medical care can already be delivered; where it is available it has ceased to be viewed as a desire but is now held to be a need, even a need based on a natural right. Organ transplantation funded by private and public insurance programs comes close to being viewed by society as a natural need insuring any individual's right to life. Vast resources, representing a sizable percentage of a society's total economy, are required to provide such high-technology medical care. New forms of medical and surgical therapy require an ever-larger fraction of the resources of contemporary Western societies.

For some widely followed medical or surgical proce-dures an effectiveness cannot be proven. For example, the commonly performed cardiac bypass procedures, where diseased coronary vessels are replaced, have not been proven to increase life expectancy when compared to careful management without surgery.[7] Implementation of such a costly surgical program for a minority of people gives high priority to providing their wants even though in a given economy, whether local or worldwide, it uses up resources for meeting more universal needs.

Where heart disease is the leading cause of death, decreases in its incidence have resulted less from innova-tions in medical care than from changes in the cultural habits of members of that society.[8] Important risk factors contribute to heart disease as a leading cause of death, yet our society has been reluctant to take an active role in reducing or eliminating those risk factors. Wealthy societ-ies like ours choose to provide high-tech health care instead.

Most societies are unable to provide such health care; they cannot even satisfy basic nutritional needs so their members do not develop the essential biological constitu-tion characterizing good health. Where biological devel-opment is retarded in a nutritionally deprived individual, the natural right to develop the human potentiality is lost forever.[9]

An almost nonexistent priority is pursuit of both re-search and provision of care for the leading causes of death in the world, certain diseases almost nonexistent in wealthy Western nations. Apart from dysenteries that cause death in children under five years old, the primary causes of death worldwide are diseases many people in the Western world have never heard of: parasitic diseases like malaria, schistosomiasis, trypanosomiasis, leishmaniasis, and filariasis. Such diseases do not attract research sup-port in the Western world; consequently most of humanity lives far short of its innate potential. One might almost say that scientists need not seek fossils for evidence of subhuman species linked to humans, because we are creating a subhuman variety of individuals anew in the world today.

For the Scientist: Truth Is in the Mind of the Beholder

Should scientists be committed to serving the interests of the public that supports science? Scientists, committed to objective, universal and value-neutral truth, condemn the intrusion of subjective, value-laden factors as being destructive to their efforts to achieve the designated goal. Would it be possible for humanity to use science primarily for solving worldwide human problems? Not easily, evidently:

> To try to commit research to a more humanitarian programme would actually be to subvert the objectivity, the intellectual integrity, the scientific character, of science. It would actually go against the real interests of humanity![10]

Science is clearly in a dilemma. Where is its promise?

Science Fiction Entertains

Confidence in science has encouraged a significant expenditure of resources on theories with no hope of serving the interests or needs of the public. Financial support has been given to ventures to detect or contact extraterrestrial life; to construct instruments to measure gravity waves, the existence of which is hypothetical; to build high energy accelerators to discover the most fundamental particles of matter; and to launch space telescopes to learn more about the universe. Such adventures are portrayed to funding agencies as important to the interests of ordinary voters.

Many believe the universe to have sufficient plan, design, order, and harmony for producing life like ours elsewhere on other planets. (Actually, the concept of human life elsewhere is not new; in 70 B.C. Lucretius believed that combinations of atoms happen elsewhere in the universe to make worlds such as ours.[11] Believing scientists convinced the United States Government to spend over seven million dollars to search for signs of intelligence coming from sources outside our planet. A Pioneer spacecraft was launched with information on our planet that included pictures of male and female humans, a phonograph with recordings, schematics of our solar system, and symbolic depiction of transformation of the hydrogen atom when the axis of spin of its electron is changed.[12] Contact with assumed extraterrestial beings is being attempted with radiotelescopes.

All of these attempted contacts are based on the premise that life elsewhere will be similar enough to us that our messages will be understood. Scientists and public alike "eagerly" await such contact even without the slightest shred of scientific evidence for human-like life anywhere else in the universe. If, however, the universe is indeterminate, existing life elsewhere will not resemble us since:

> The evolutionary path that led to the appearance of *Homo sapiens sapiens* was so tortuous and complex that there is no possibility of its being repeated in other worlds. They will look like nothing we have ever seen before.[13]

In another adventure the space telescope is said to be important in the study of black holes, which may hold the key to the universe's ultimate destiny.[14] Information gained would be useful for commenting further on whether the universe is expanding or contracting. Theoretical physicists believe that we might learn enough to revise certain laws of physics, perhaps to prevent an extinction of the universe billions of years from now.[15] But if the universe were found to be unstable, changing in one direction or another, how could that finding be important to human beings when any effects from an expansion or contraction would occur millions of years from now? People are not doing what is necessary for insuring survival on this planet over the short period of the next century. What the universe becomes in the very distant future is not even a significant academic question if human beings cease to exist as a form of life.

Another research proposal is to develop space technology for inhabiting other planets when our present one becomes uninhabitable, although we know of no other planets capable of supporting life. Why should resources be spent on seeking such a place rather than on maintaining the only place known to be capable of supporting life?

Although ordinary citizens understand virtually nothing about efforts to describe the nature of the most fundamental particles of matter, vast resources are spent for such research. In an era of huge budget deficits, however, governmental support agencies do seem to be abandoning the kinds of basic research showing no promise for answering questions of immediate importance to human beings. There seems to be "a widely held suspicion of high-energy physics as a luxury whose fruits do little to nourish the public at large, or even other fields of science."[16]

Seeking Truth; It Must Be Universal

Use of the world's resources for applied science is justified, but a new priority must be emphasized: the goal must be to satisfy universal human needs. That means that the desires of some people must be subordinated to the needs of all humanity. Before that can happen people must recognize and accept the purpose for science.

Science for Insuring Survival, a Natural Human Right

We have seen that the early Greeks were aware of practical applications for science (for example, in the works of Archimedes), but they generally lacked motivation for developing a scientific technology. Their lack of incentive is explained only partially by the abundance of slave labor or by any lack of financial support by the ruling society. Eventually, application of science became absolutely necessary to assure humanity's survival, which had not been so seriously jeopardized in earlier times.

Scientific technology has made survival of the human species possible into the distant future. Science cannot instruct human beings about how to live; its purpose is to show them that they *can* live and to satisfy their needs. Ignoring the purpose of science could jeopardize our continued existence.

Biologically, Humans Are Not Privileged

Science can be used to prevent overpopulation of the world and the accompanying threat of devastating new infectious diseases, as are seen with the overgrowth of any biological system. In a closed biological system, life expands, the number of living things tending to increase exponentially.[17] The rate of population growth is controlled by such factors as the availability of food and the appearance of disease. The dynamics of growth of living organisms can be studied by introducing rapidly growing forms of life into a closed system containing a finite amount of nutrients and representing a limited environment for disposal of waste products. For example, bacteria inoculated into broth in a test tube begin to multiply rapidly with an exponential growth rate. Growth continues at that rate until affected by the limiting factors of nutrient depletion and production of inhibitory byproducts, after which the rate of growth decreases (a rate that is also exponential). Bacterial growth can be graphically depicted in the same logistic curve which accurately describes the growth of the U.S. population.[18] Thus the mathematical description of human population increase is no different from that of bacteria (Figure 4). If human beings were merely a higher form of animal life with no significance in a meaningless world, there would be no design to give people greater value than other life; human beings would have no means to survive any better than penguins in Antarctica or a culture of bacteria in a test tube. But there *is* a design; human beings were given science at the right time of need to equip them for insuring their survival.

The varieties of life on earth are myriad, ranging in size from electron-microscopic viruses to Sequoia redwoods and humpback whales. All life lives in ecological niches in which the size of each population is controlled. No form of life is able to multiply wildly, overflowing its numerical limits, without being subject to forces that limit its numbers. For all subhuman forms of life, we recognize the limiting influences as essential for maintaining the tremendous richness of life throughout the world. Food supply and disease, the major factors maintaining population balance, are in the design whether described as due to evolutionary processes or to God.

The limiting forces controlling human population on an individual basis or a large scale are generally viewed as evil forces inconsistent with a world designed and operated by God. The fact that limiting forces exist is not evidence to deny the existence of God, however, if God is concerned with what happens in the world. The human biological constitution is essentially the same as in other living matter, so it would be surprising if the forces limiting biological populations had no influence on the human populations.

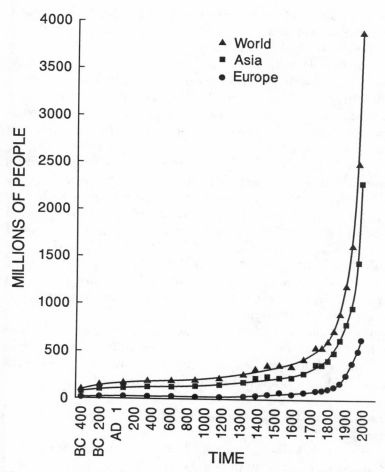

Figure 4. Populations of World, Asia, and Europe from 400 B.C. to 1975. Redrawn from data in C. McEvedy and R. Jones, Atlas of World Population History. Although there is a different beginning point for each curve, there is no significant difference between the three curves; the dynamics of growth have been the same for all three populations.

The freedom for human populations to multiply within limits could also be explained strictly by evolutionary forces, with no design or concern by God. In that view, of course, the limiting forces of food supply and disease are not evil and human populations are in no way privileged to be unaffected by forces affecting all other life.

For people to believe that human beings are privileged would be a value with no scientific basis. Since science must be value-free to be universal and objective, a scientist would consider such a belief irrational, or perhaps transcendental (with no empirical support). The value of being privileged cannot be derived from science or human reason.

Freedom within limits, a general principle for all life, is not evil; it is necessary to maintain the beautiful diversity and order that have evolved throughout the world. The requirement to maintain a balance has been implemented in human populations by famine, disease, accidents and wars. "Charles Darwin noted in *The Descent of Man* that infanticide has been 'probably the most important of all' checks on population growth throughout most of human history. Infanticide continues to be practiced and it is interesting that it has been documented so well in the highly civilized societies of the Western world."[19] Constraints which operate to control human population could be judged as evil and incomprehensible in a universe designed and operated by a concerned God.

Evil and Suffering; Humanity's Choice, Not God's

Although subject biologically to the limiting forces exerted on all life, human beings are allowed the freedom to control those limiting forces so they interfere minimally with the satisfaction of natural needs. Our freedom within limits includes a freedom of choice about whether to control the activity of forces limiting human population. Failure to exercise that freedom permits the "evil" that such forces can promote in a God-concerned universe. The evil appears because of intentional default by human beings. When people neglect to exercise their freedom of choice, a gift of their human potentiality, a concerned God cannot be blamed for the evil that ensues.

The evil appears because people do not use their human potentiality to control the growth of their numbers or to control the forces limiting population size. Controlling human population growth intentionally would minimize expression of the natural limiting forces. Controlling factors designed to limit population would avoid their devastating effects.

The great diversity of life on earth would have been impossible without a design to limit the growth of each form of life. The design is essential for an evolutionary explanation of the development of life; it is also a necessary condition for God to maintain the rich diversity of creation.

Limits to the Human Uniqueness

Human beings are not biologically different in significant respects from other primates or even other mammals. The human uniqueness lies elsewhere; it is in the potentiality for developing means to exercise control over limiting forces, and that potentiality extends to a freedom to decide whether or not to use those means. Humanity gained science and technology as one tool for controlling the limiting forces determining survival of the human species.

Cultural Imprints to Control Human Population

Culture is essential to human beings partly to assist them now in controlling the size of their population. For a human population some of the factors determining growth rates are age, sex, marital state, occupation, education, and religious beliefs.[20]

In the United States a decline in the population's growth
rate is seen like that of bacteria in a test tube, but the
factors responsible for the decrease are totally different.
Among those factors are higher standard of living; lower
illiteracy, increased lengths of time for education and
training for a subsequent career; greater freedom of
choice in pursuing career goals rather than limiting interests to that of a parental role; greater opportunity to
satisfy wants or desires that are gained at the expense of a
commitment to child-raising; and less importance for
meaning or fulfillment as a parent.[21] Similar declines in
growth rate are seen in populations of other societies with
a high standard of living.

Stabilization of population numbers is considered
possible only when many factors contributing to a high
standard of living are achieved.[22] If populations cannot
improve their standard of living until they achieve
industrialization, then no society could control its population growth until it industrialized. In that case, Third
World countries could not be expected to change, since
countries with the poorest standards of living have
virtually no hope for industrialization; small scale farming
is seen as a more realistic means for improving the human
condition.[23]

Standards of living are relative; the improvements
sought are usually intended to go beyond meeting people's
natural needs to satisfy some human wants or desires.
Attempts to achieve such a goal universally have little
chance of succeeding. Even if industrialization were the
answer to improving a society's standard of living, with the
hope that stabilization of population would follow, there
would not be time for that to happen before the pressures
of increasing population would reach explosive proportions. Science has given humanity the possibility of stabilizing world population without relying on industrialization. To use science and technology for achieving that goal

would manifest our human freedom of choice; thus people have the option to exercise control of their destiny and they need not view all manifestations of their destiny as merely determined by a design of God.

Why did Human Population Grow?

The rate of growth of human populations remained very low and stable during early history. With the development of agriculture around 5000 B.C. the rate of growth began to accelerate. The acceleration formed a primary cycle of growth seen first in the Near East and later in Europe and China, with the first cycle lasting about 6000 years.[24] The primary cycle did not begin until later for societies in the Americas, Japan, and sub-Saharan Africa. The end of the primary cycles in the oldest classical societies was associated with a decline in their cultures after an evolution and rise to a zenith. During the primary cycle, society in a sense played out its full history, as that society's numbers multiplied up to and beyond the optimum for the technology of the time.[25]

The next increase in population growth is termed the medieval cycle, which appeared at the same time in both Europe and China.[26] This cycle appeared with the emergence of a feudal society. Transition to aristocracies may have resulted from technological advances or other factors also responsible for a change in population growth rate, but the reasons behind population growth cycles are not well understood. Certainly, some pauses were caused by wars, famine, and epidemics.

The next cycle, that of modernization, began in fifteenth-century Europe and a little later in China.[27] Technological advances in the construction of ships made it possible to cross the major oceans; the development of firearms allowed societies to dominate and colonize newly discovered parts of the world. These early beginnings culminated in the agricultural and industrial revolutions of the eighteenth and nineteenth centuries. Since then, cycling of population growth rate is less apparent, with rates of increase being maintained at high levels.

Arguments that agricultural and industrial advances were essential for the growth rates of the latter part of the modernization cycle are not convincing. Those arguments maintain that the rate of human population growth throughout history is determined by technological advances. Yet the rapid population increases in China during the current century were not promoted or sustained by agricultural and industrial advances. Agricultural methods employed in China had changed little over many centuries. If food was a limiting factor (which it was at times, as shown by population decimations due to famine), there should have been a decrease in the rate of population growth of China.

Currently, Africa has the highest population growth in the world. But this high growth rate is associated with an African food production that is less than it was twenty years earlier; not surprisingly, this has resulted in widespread famine.[28] Thus the rate of population growth remains high despite reduced food supplies, persisting until a catastrophe by famine. Such an outcome appears sooner when human death rates are reduced by modern medical technology.

Human Population Control; by People or by Nature?

Population growth rates do not seem to decrease spontaneously without human intervention. When people do not restrain their own population growth, their numbers are controlled by forces dictated by the world's design. Throughout history, populations have suffered catastrophes of wars and other forms of violence, famine, and epidemics of infectious diseases. The twentieth century has seen these three catastrophic factors acting in concert. Famine and disease epidemics are invariable partners in the decimation of human numbers. With science's technology and freedom of choice, humanity could dedicate the resources and could design the programs for controlling population growth while providing the needs of every person. To be successful, the endeavors must be on a global basis; attempts at merely a regional level are not likely to succeed.

The Universal Unbalance: No Unity, Order, or Beauty

In countries with a high standard of living, competition is among people trying to satisfy their wants or desires; their needs are usually met. Only certain individuals will "fall through the safety net." They include people with major physical or mental handicaps, those with no desire to work, and small children and the elderly who have no one to provide for them.

Most of the world does not enjoy a reasonable standard of living. Overall, the Third World has a population growing rapidly at an annual rate of 2.4 percent.[29] That world includes some societies (in Central America, for example) where agricultural and technological advancements of the twentieth century are available and others (such as parts of Africa) where they are not. Regardless of the availability of technology, in most of the Third World adequate nutrition is a basic need that is not met. Any standard of living is unreasonable if people are unable to secure their natural right of adequate nutrition.

Should we expect continuing progress in agricultural productivity to begin solving the problems of inadequate nutrition? Several factors make it difficult for improvement in agricultural technology to solve the problem. The combination of population pressure and deterioration of usable land has put fertile land at a premium. In some

African countries deserts are spreading, forests are de-
nuded, and land is overgrazed, all of which are leading to
rapid deterioration of the country's agricultural capability.[30] Competition to meet the absolute need of food,
which may have been an excuse for war in the history of
humanity, now foments new types of violence.

Competing to Survive

Although immigration was once desirable to provide the
necessary labor force in underpopulated societies, there is
now pressure to limit the entry of outsiders into more
developed societies.[31] Immigrants from Southeast Asia to
the United States have experienced discrimination and
violence from people perceiving their presence as an
unwarranted competition for employment in work necessary for them to most importantly satisfy their wants and
desires. In India many immigrants from Bangladesh seeking work for a better life were killed. In Vietnam and
Nigeria the government of the majority carried out mass
evictions of peoples with different national origins. In
Indonesia hundreds of thousands of people with different
ideology were killed by the Muslim majority of that
country. Thus, the pressures of deprived population densities and the competitions they generate lead to violence.
    Another potential disruption of the deprived society
comes from young adult males. The most volatile age
group in most populations is that comprising fifteen- to
nineteen-year-olds. With one-half of the Third World
population being sixteen years of age or less, a majority of
the population is in or at the brink of entering the most
volatile age.[31] Where deteriorating conditions have resulted in failure to imprint these young adults with culture's stabilizing influence, the potential for violence is
great, especially in cities with the highest population
densities.

Most cities in the world, especially in Third World countries, are experiencing growth due to a movement of people from rural areas.[32] People move to the city in the hope of improving the quality of their lives, a hope built on the wealth to be seen in every city of the world. Overgrowth of cities outstrips their capacity to provide services, however, including delivery of people's basic needs. With that migration contributing to an already high population density, cities in even some technologically advanced societies experience crowding. The slums created breed crime and tension, forming a fertile ground for the activities of extremists and a potential for greater violence. In addition to lacking the basic needs of food, shelter, sanitation, and health care, slum dwellers are also deprived of freedom from violence and fear.

Control of Human Population Size by Disease

Disease will contribute to controlling human population when certain limits of its size are approached and surpassed and when basic needs of shelter and nutrition are not satisfied. When poor people crowd into slums, disease is likely to seriously affect newcomers. Following universal exposure to infectious diseases, immunity is built up, causing mortality and morbidity rates to decline. Populations acquire resistance to infectious diseases so that if exposed and infected, a less severe form is manifested in comparison to the severity of illness seen in populations who have never encountered the disease.

In addition to undernutrition interfering with economic development in poor societies, it magnifies the consequences of most diseases. Acquired and inherited levels of immunity to all varieties of infectious and parasitic diseases are reduced with nutritional debilitation; people are more susceptible to such diseases in deprived societies. There is an even greater potential for devastating effects of nutritional deprivation on populations when we consider the threat posed by new diseases.

Control of Population Densities by Viral Diseases

Viral diseases have the greatest potential for devastating any susceptible species; that potential has been documented for a variety of viruses infecting domestic and wild animals. Viruses are able to adapt or evolve into variants which in reality represent new viruses never seen before; some of them attack mammalian cells in completely new ways. Evidence suggests that the appearance of new viruses correlates well with the introduction of "factory farming" practices to increase population density.[33]

Some viruses have appeared with the ability to wipe out entire animal populations, something no human viruses have shown a capability of doing, at least to the present time.[34] For example, the Newcastle virus first recognized in Indonesia in 1926 can destroy an entire flock of poultry; the hog cholera virus first seen in 1833 is deadly for swine populations; and the myxoma virus, present for hundreds of years, continues to be capable of destroying entire populations of rabbits.

Swine influenza (appearing in 1918), parvovirus infection (appearing in 1966), transmissible gastroenteritis (appearing in 1946), and rotavirus infection (appearing in 1975) are examples of viral infections that have appeared with the intensification of production and increased population density; all are capable of infecting entire herds of swine and in some cases have wiped them out. With these animal diseases, the viral agents spread rapidly from affected to clean populations; the severity of the epidemic depends on the intensity of local livestock production, the amount of movement of animals in and out of the population, and the extent and complexity of the disease.

It is absolutely predictable that new viral disease agents never before recognized will appear and infect animals. This should also be anticipated for viruses producing disease in humans. Some new viruses are infective for more than one species of animal and they acquire a greater virulence when infecting new hosts.[35] Other viruses present for years in a wild host show a virulence for a domestic species of animal that they come to infect as a new pathogen.

Viruses have evolved the capacity to infect any living cell in the mammalian body; moreover, all forms of life, plant or animal, can be infected by viruses. Evolutionary processes continue to change the biological properties of viruses. Such changes can be accelerated by intensification of animal or human population density. With greater opportunity for rapid spread and multiplication there is a greater chance for viral change through the evolutionary process.[36]

Scientists know how to manage most of the threats posed by human and animal viruses after learning how to isolate and grow the organisms outside the host and then to use cultures of those viruses for developing vaccines. That approach has been successful in almost eradicating the viral disease smallpox from the world. The solution to that problem was relatively easy, however, compared to challenges appearing in recent years. The new problems concern attempts to protect living populations against some of the new and rapidly changing viruses.

New viral diseases are often caused by viruses that elude all scientific attempts for their isolation and cultivation, the first requirements for solving the problems they cause.

Some of the more threatening new viruses have the potential to attack the body's normal immune system. Animals or humans showing signs of disease after infection by such a virus do not recover; they eventually die from the infection. The virus destroys the only protective mechanism that living tissues possess for responding to an infection by any agent.

Most frustrating to biomedical scientists are viruses that evolve so rapidly that what characterizes the virus today can change tomorrow without impairing the virus's ability to cause disease. An antigen is a substance, usually a protein, that stimulates the immune system to produce antibodies in an immune response directed against the antigen. The immune system mounts its defense against infectious agents by reacting to the antigenic substance that make up their structure. Scientists are unable to keep pace with a virus that can change its antigenic makeup rapidly, making it impossible to produce vaccines to stimulate protective immunity. The ability of human viruses to change is illustrated by the influenza viruses which spread rapidly through populations. Before the "flu" virus leaves one population to move on to another its antigenicity may have changed, so that a vaccine made from the initial form of the virus may not protect people against infection by the transformed virus.

Genetic determinants of a virus's potency for producing disease and of its antigenicity can change and be expressed only when a new generation is produced. It is evident that such change occurs more easily with forms of life reproducing over very short intervals of time, as is true for most viruses. With viruses replicating as rapidly as any form of living matter, there should be, over a short period of time, a greater array of possibilities for changes in viruses compared with all other forms of life.

A Viral Agent for Ultimate Population Control

The most devastating viral agent imaginable is one that would (a) have the ability to infect cells of the immune system; (b) change its antigenicity frequently enough to preclude production and use of vaccines affording effective immunity; and (c) acquire a great virulence or potency to infect, thereby enabling it to spread rapidly (as with influenza in humans and Newcastle disease in animals). A virus has appeared in acquired immune deficiency syndrome (AIDS) with the first two of these properties; any comparable virus acquiring the last property could totally devastate populations of susceptible humans or animals.

The possiblity for such a viral agent is less of a "wild idea" than the concepts of the big bang and quarks. Evolutionists realize that this type of virus could appear if the virus is given enough chance for producing many variations, and expect that to be only a matter of time. It is possible that such viruses operated to decimate animal populations at times in the past, which might have represented a purposeful tool or response in the evolutionary process. One can see natural processes in the design of the universe for controlling human and animal populations; if people neglect or abrogate their potentiality to control population themselves, it will happen by natural processes.

More than Science Is Needed

Science-based technology provides one means for controlling population growth rates by devising methods to prevent conception. Such accomplishments are ineffective, however, unless value can be shown for their implementation. Intensive education of masses of people is necessary for them to recognize that it is in the best interest of their society and of future generations to control population numbers now.

Technological achievements to increase world food production have also helped minimize devastations of expanding populations. Except in one or two societies, enough food is currently produced in the world to prevent starvation. For many people, however, the types of foods produced are unable to provide a nutritionally balanced diet. In most of the world a low ratio of protein to carbohydrates contributes to less than adequate growth and an increased risk for infectious diseases. The technology for improving food production may have peaked, so that future crises may not be averted by any new successes in dramatically increasing world food production. Agricultural technology also requires education of farmers on a worldwide basis; beyond science's accomplishments must be a consideration for value.

Fulfilling the Promise

The single importance of science and technology is for controlling the natural factors threatening humanity's continuing existence at a time when its numbers are increasing rapidly. Science-based technology enables people to control their numbers so they will not be controlled by nature's forces. It shows how to minimize the effects of old and new diseases, which ally with population explosions and nutritional deprivations to create a natural means for population control. Are we fulfilling the promise?

Chapter 9

Humanizing the Species

Technology Does Not Civilize Human Beings

Western society has an unwavering faith that technology will in a matter of time solve global problems. Other than those for physical survival the major problems facing people are social, not scientific. Scientific advances seldom solve social problems beyond the extent indicated here in solving problems of unchecked population growth. Primitive societies that were (and some that are) technologically in the dark ages show evidence of a highly developed degree of social advancement.[1] On acquiring technological proficiency they do not advance to a more ideal society. Societies unsuccessful in developing their own technology or indifferent to any value for acquiring technology are not necessarily unsuccessful in the ordering of their human relations. In fact, it appears that societies directing most of their efforts to economical and technical advancement may have little energy left for construction of a stable society.[2]

There are many instances where great stability of a society is associated with little technical advancement. For example, Eskimos who have not enjoyed technological developments until very recent times, expending only enough energy to meet their bare requirements, are recognized as having developed a society with valuable traits not seen in "more civilized" societies.[3] Those traits include a surprising ability to control the urge to satisfy one's desires, associated with an unusual kindness to others and responsiveness to the needs of others. The Australian Aborigines, considered more primitive than Eskimos and sometimes depicted as remnants of the Stone Age, have a social organization, with marriage rules, taboos, philosophy, and ceremonial life in many respects no less complicated than our own.[4]

Modern Western people do not always see any value in the societies of technologically primitive people. Yet "the tone of human relationships can be more satisfactory among people who feed on leaves and grass than among people who have reached the last perfection of manipulatory efficiency."[5] Thus, while scientific technology has progressed so that the population problem can be approached with a scientific capability for its solution, the sociality necessary to implement any solution may have regressed to such a point that little can be accomplished.

The Ultimate in Human Sociality

If there were continued progression in what is held to be an evolutionary descent of human sociality from animal sociality, consisting of a long-drawn-out series of gradual adjustments, the ultimate in human sociality would begin to appear universally. That would be seen in food sharing, something seen only in parent-offspring relations in animal society, even among the so-called highest form of nonhuman life in primates. The cultural trait of food sharing contributes to human society and is the single most important attribute demonstrating self-denial of desires and wants. Food sharing contrasts with the self-indulgence characteristic of animal societies.[6] Food sharing sociality, is not achieved with scientific technology and should not be sought there. Food sharing is commonplace in some technologically primitive societies, but may not be a dominant value in the most technologically advanced societies.

Food sharing has a priority hardly related to the wealth of any nation. For example, as percentage of gross national product, United States foreign aid to help the hungry is half that of West Germany and one-seventh that of Norway. Aid given by the United States is determined by mutuality of strategic concerns, relations of the government to those with needs, and concerns about whether such aid in the form of food might disrupt the agricultural economy of the recipient society or perhaps of another country. Many individuals in the United States do not believe that our government should increase the level of foreign aid, the notion being that we should take care of our own "needs" first. The Western world cannot claim realization of the ultimate in human sociality.

What does characterize human relationships in technologically advanced societies and what is most acceptable for humanity's existence? Is it where evolutionary advancement for a society is manifested by accumulation of material possessions by the recklessly selfish, and where that becomes the most admired and emulated trait? What passes for civilization for many Western people is far from the ideal human relationships of society. In hunting and gathering societies the most admired human attributes are generous giving and sharing, not the accumulation of possessions.[7] If food sharing is an ultimate human achievement, the culture of the most "highly civilized" people on earth must have regressed. But the ideal of human society can be regained:

> Man . . . never rises above himself with more brilliance than when he subdues his own nature to the point of making it follow a way contrary to the one it would spontaneously take. By this, he distinguishes himself from all the other

creatures who follow blindly wherever pleasure calls them; by this, he makes a place apart for himself in the world.[8]

Technology and Cultural Regression

The road back to giving and sharing on a universal basis seems unattractive in societies placing increasing importance on accumulation. The difficulty in taking that route to solve human problems is not with inadequate scientific technology. Modern Western society experiences incidences of family instability, crime, suicide, alcohol and drug-related disease, mental disease, terrorism, and warfare that are unusual in poor, nonindustrialized, "primitive" societies. The more technologically advanced societies frequently show a permissiveness related to weak or ineffective social control.[9]

Evolutionary Development for Human Potentiality and Sociality

Beyond biological evolution, is there also an evolutionary process by which human beings have been transformed from an animal condition of self-preference (i.e., automatic obedience to body-based urges) into humans with a culture that brings biological urges under control so that higher values can influence their conduct? Scientists cannot explain why one specific primate gained the potential for cultural evolution, except that the subhuman primate was in a unique position to build on the genetic potential for brain function accumulated over the previous millions of years.[10]

The basic features of evolutionary theory include subtle transitions from one degree of complexity of structure and function to the next. Different degrees of the human potentiality and sociality should have appeared in the animal kingdom just as the different biological forms appeared in the evolutionary descent to human beings. No such gradual changes are found for the mental processes unique to human beings; no trace of the human potentiality or sociality exists in any other form of life. The animal mental processes and sociality (what there is of it in even the highest nonhuman primates) are qualitatively and quantitatively different from that seen in even the most primitive human societies.[11] Hardly any connection can be seen between the mental uniqueness of the highest animal and the most primitive human society. The idea that very gradual changes eventually resulted in differences of the kind seen between animal and human mental processes is a scientific theory for which there is no evidence. Whether or not it is the product of any evolutionary process, the social outcome is totally different for primates and human beings. Why did those traits evolve in only one form of life?

There remains a tremendous gap between the sociality humans are capable of and that for any animals. Scientists are unable to close this gap, and it is hard to imagine that "fossil behaviors" could be found to fill in any missing links. That does not mean that scientists have not tried. Speculations about the development of human sociality have been based on archeological discoveries; they include:

> . . . suggest that these primitive hunters may have preferred to bring prey back to the cave and share the meat with others in the community rather than to consume it where it was killed . . . [why?] . . . . [it] may have been difficult for women because of physiological limitations (such as pregnancy and child rearing). It is thus possible to speculate that the hunting behavior of Peking Man may have caused or contributed to the sexual division of labor within the group.[12]

Others have different suggestions based on the same observations:

> There is no reason to believe that Peking Man even had a
> cave home, much less hunted . . . the animal and human
> bones [found] were dragged into the cave by hyenas . . . .
> The ashes on the cave floor may have not come from
> cooks' hearths but from forest fires; perhaps burned or
> smoldering branches fell through cracks in the cave . . . .
> Much of the "ash," . . . is actually owl droppings.[13]

(What a difference in these two stories; such is the under-determination of some theories.)

The evolutionary explanation for lack of visible progress in human cultural development is that the process has not had enough time to work on the raw material of the human animal.[14] The examples of the Stone Age remnants and Eskimos described earlier show that, sheltered from the mainstream of civilization's progress, they maintained a sociality closer to the ideal than modern cultures. That speaks against any evolutionary process for culture.

Scientists Explaining the Human Potentiality

Many scientists anticipate that an artificial human intelligence or consciousness can be created, believing that all human mental processes can be formalized to strings of symbols which can be duplicated with a machine.[15] Mitchell Waldrop writes:

> In 1859 Charles Darwin published *On the Origin of Species*
> and made *Homo sapiens* one animal among many—and
> gave us a far less exalted view of our own significance. And
> now the creation of an artificial consciousness could
> change our very ideas of "self."[16]

Human mental processes of logic and language structure have been argued as being formalizable by such philosophers as Plato, Leibniz, Russell, Whitehead, and Chomsky, but the basic premise has not been proven. The knowledge, intuition, insight, inspiration, and imagination of human intelligence must remain even more transcendent, being more difficult to formalize than reasoning, logic, grammar, and language structure.

A Scientific Basis for Human Behavior

Nothing in the evolutionary biological process indicates why humans should have evolved into social beings showing love, empathy, and mutual concern for a social and cultural order. Yet people do show characteristics supporting the idea that "survival of the fittest" is a continuing goal of evolution. In addition to the biological competition necessary in Darwinian evolution, the human traits of anger, greed, and jealousy are natural emotions with obvious advantages in a society where a large number of strangers must compete for a small number of positions of dignity, status, and economic security.[17] Survival of the fittest drives people to maintain a self-interest; we suffer when we appear too cooperative, too loyal, too altruistic, or too reticent in protesting frustrations or advantages seized by others. But most people are not completely self-serving and not totally subjective in their perceptions of humanity's needs and interests.

People also reach out with concern for fellow human beings, manifesting love, sympathy, and self-sacrifice for some, and even enduring pain or hardship for others. Social theory tries to explain these human activities by objective, value-neutral, descriptive laws, viewing humanity as a purely natural phenomenon, the ultimate product of evolutionary processes. But a value-neutral science of human beings in society fails because it is unrealizable and self-deceptive and its theories are more underdetermined than those of the natural sciences.[18]

Characteristics such as empathy and love are sometimes claimed to be innate human attributes—that is, they are inborn realizations of the evolutionary descent of humans from animals.[19] But such characteristics are totally inconsistent with Darwinians attributing human evolution to chance or random variation and natural selection so that selfish genes can insure their survival and propagation. Objective, value-neutral explanations fail to explain the appearance of value-laden attributes incompatible with evolutionary criteria for human descent from animals.

Human Potentiality: A Creative Design

The human potentiality is part of a creative design. That includes both a design for culture which was provided to fulfill a need and a design for science which was given to fulfill another need; both gifts are necessary for achieving a meaningful existence. This is the only explanation other than an evolutionary one for the unique human sociality. As people began to deny any explanation by design and sought scientific explanations for their potentiality, they regressed from the cultural ideal and adopted varying degrees of self-interest, self-indulgence, and self-preservation. That consequence should be sufficiently convincing to abandon a scientific theory counterproductive to progress in the development of the human potentiality.

If human uniqueness is beyond any objective scientific examination, where did human attributes come from?

Is There an Innate Goodness?

How do we find good in ourselves in a meaningless, indifferent universe? Are people innately good? Are we endowed with some degree of innate goodness which can by constant effort be eventually perfected? For some people a humanity with innate goodness lies in the essence of human beings. "For this is humanism: mediating and caring, that man be human and not inhumane, 'inhuman,' that is, outside his essence."[20] If value-determined desirable traits are innate, individuals should show evidence of them from birth. It is almost impossible for scientists to determine what is innate and what is learned in the very young human infant, however. Beginning at the time of birth, the infant is placed in an environment with adult humans who immediately begin to mold its responses. In contrast to animals, human beings have essentially no instincts—no innate preformed patterns of behavior.[21] All human responses to the environment must be learned; empathy and love are not innate as a part of an inborn "goodness."

The lesson from accidental raising of human infants as animals is that unique human qualities such as language and intellect do not appear. Some feral individuals did not even develop the human postural characteristics for sitting or supporting themselves on their feet.[22] Thus, even that behavior is not instinctive but is acquired or learned from contact with society. Upright human posture is acquired from culture; feral humans ambulate on all four limbs as do animals; when standing their posture is stooped.[23] Only animal sounds are made by feral humans. The unique human achievement of language is culturally learned by every human being; only the potentiality for language is inherited.

It might be expected that the sexual behavior of feral humans would be as intense as that of animals. Actually, post-puberty feral humans show little intensity in their sexual drive and what there is remains inchoate and unfocused.[24] Thus, for the expression of even such a basic drive as the sexual one, the feral human seems not to have instincts or innate responses; the realm of human culture is necessary for people to manifest the full expression of their sexual identity.

The feral child shows no sociality and looks on people as

objects just as we would look on animals.[25] Such children may show hatred for or disinterest in humans of a comparable age, and they show a preference for solitude. All the activities and emotions of isolated or feral children are concentrated on satisfying their natural needs of eating, drinking, sleeping, and survival. With all its responses consistent with being a perfect egoist, the feral child appropriates and keeps whatever is found, shows neither affection nor pity, and will sometimes attack anyone who approaches.

Although such individuals were hardly different from animals when found, they were still human beings. They were born with a capacity for developing human characteristics. At birth, the environment directs that development; later, when the individual is able to choose, habits are adopted, enabling human beings to determine what they become. No animal has any potential to choose what it will become; that is genetically determined before birth.

The human potentiality is not the translation of a genetic code; it is no more than a promise that a creature can develop into a human being. The potentiality is fulfilled as individuals receive the stamp of society, after which they can exercise the uniquely human freedom for choosing further development. The human potentiality allows us more than any other form of life to enlarge our range of freedom by recognizing and understanding the limits to that freedom. If that freedom of activity were merely an expression of an evolutionary random variation, it would have no restricting limits and the chance occurrences of its expression would represent another manifestation of the meaninglessness of indeterminism.

Individuals Doing It "My Way"

People do not wish to believe that there are limits to their potential, however, so that with greater personal commitment of the intellectual and the emotional working together as one, they believe that the "ascent of man" can continue. The historical events of the last two thousand years do not support the idea of unlimited progress and the events of this century deny it. Human beings from the time of Genesis have wanted to believe that "knowledge is our destiny" without which we shall not exist; we can gain an understanding of ourselves if we devote ourselves to it.

If it is true that little substantial has been added to philosophy in twenty-five centuries,[26] it seems even less likely that knowledge through science will add anything new. The scientist-philosopher is unduly optimistic about the methods of science, and of physics in particular, leading to truth.

It is a myth that scientific knowledge can create a new human being capable of living in harmony with others. Even if social theory gains the status of scientific knowledge, objective study of human nature is not likely to give moral philosophy a sound factual basis for the emergence of a new humanity. The study of human nature by Freud put the animal back into humans; more precisely, Freud claimed that the animal never completely left during the evolutionary descent. With the animal still in us we can continue doing it "my way."

"Let Freedom Ring"

Human beings seek freedom, guaranteed as a natural or God-given right. The shadow of freedom is insecurity. Human nature bears in it the dark seed of discord; let people be free and that seed of discord will germinate and blossom, not the flower of harmony that is falsely attributed to the laws of nature. Freedom does not make people kind and unselfish; it has the opposite effect. True and complete freedom isolates humans from one another; under those circumstances the self-preference and self-indulgence inherent in the newborn appears anew. If freedom's shadow is insecurity, order's light is security.[27] Freedom inversely follows order.

Excessive freedom (in the form of ultimate privacy or isolation) and excessive denial of freedom (loss of any privacy, with constant intrusion by others) are equally undesirable.[28] But the propensity to greater freedom is more harmful than the propensity to less freedom. Internalization of culture's values gradually removes the freedoms of isolated humans, giving a security that is manifested in cultures as a minimum of fears, frustrations, repressions, and ill-adjustments common to animal life; and a maximum of tranquility and fulfillment.

Prosperity and Social Progress

The degree of prosperity enjoyed by societies affects the level of social control. People living under conditions of difficult existence, manifested by long hours of work and satisfaction of only essential needs, are subjected to a form of social control leading to growth in sociality.[29] Such conditions induce and support a strict discipline that is often cheerfully accepted. Societies constrained by these conditions tend to survive longer than comparable ones able to accumulate wealth. In fact, in the older, more religious, and more ascetic societies, acquisition of wealth was a greater threat to their continuing survival as a culture or society than deprivation or even poverty. In such circumstances it is the poor who are contented with their lot, the rich who are not satisfied. Sociality survives and grows with the self-discipline imposed by deprivation of all but essential needs; but beware the self-preference, self-indulgence, and greed of prosperity and freedom.

The goal of science should not be to produce a universal prosperity based on the values of Western cultures. Those values are excessive, based on desires and wants instead of needs. For emerging societies and nations such prosperity would be destructive for sociality and its human benefits. Science has the opportunity to concern itself with the totally opposite cause of social problems, that of excessive poverty. It should be possible to treat poverty successfully without striving to give the world a degree of civilization damaging for cultures and societies.

Toward Ultimate Sociality, the Only Universal Truth

The internalization of culture, suppressing the infant's natural desires in favor of human cultural ideals, is best achieved when people are able to reduce their wants and desires and focus primarily on their essential needs. That limits consumption of all kinds of resources, redirecting efforts to see that people universally shall realize their natural needs, something implemented when human beings recognize universal food sharing as the ultimate in sociality and cultural growth. The disease of many people's deprivation begins to be treated by managing the disease of a few people's surfeit, not simply because more material wealth is spread out but because of a change in values. The body of an individual that could experience only its own hunger pangs ascends now to its full potentiality and begins to experience the hunger of others as deeply as its own, and the needs of both are fully satisfied.[30]

With Sociality Only Skin Deep

Humanity exists continually in a precarious position, as illustrated by the relative ease with which people in a mob or riot spontaneously regress to a non-social or pre-social condition, giving the appearance that any internalization of culture is completely lost.[31]

We have noted that in the major part of the world, half of the population is less than sixteen years old. With the young not internalizing culture to the degree found in adults, animal behavior remains closer to the surface, with a greater potential for expression. It is important for members of the "most advanced" civilizations of our time to recognize that:

> There is no society, however materially, mentally, and even morally advanced, in the depth of which there does not remain a natural and innate trait which, if it escapes control, is a moral threat to society, to the common life.[32]

Control has broken down with increasing frequency in our technologically progressing societies.

It is also important to recognize that people do not become privileged in any way culturally because of vast scientific and technologic achievements in their society:

> There is no society, however materially primitive and mentally unsophisticated, which does not achieve the necessary minimum in the moralization of individual conduct, so long as its cultural traditions are intact, chief among which is the established system of social control.[32]

The moralization of individual conduct is diminished when people believe that science can provide the framework for a system of social control.

The "Sin of Adam"

Without the imprint of society, humans born today are like isolated people and will exhibit a behavior that has not developed beyond that of animals. Stripped of society, human beings do not develop; they do not progress beyond an animal state despite their human potentiality.

Judeo-Christian religion embraces a belief that, since the time of the fall of Adam, humans are born "in sin". Human infants are hardly innocent creatures who only subsequently learn to "sin." Babies are born with a measureless greed that demands immediate gratification, coupled with all kinds of means for exploitation to gain satisfaction. Nonbelievers have trouble accepting that humans are created with the "sin of Adam." The fact is that they are created with an animal behavior; only by means of a culture or social control internalized in their youth do they gain any innocence or freedom from sin.

Sociality is in no way innate or inherited; it must be learned or acquired anew by every individual. "Human decency" is not inherent in people; it must be acquired from society, assuming that a particular society has it to offer. If culture does not have that to give:

> Satan tells Adam that his quest for good community is hopeless because man (Thou) must obey the animal within thee, so long as matter lasts, so long shall I prevail on earth.[33]

Thus, people stand astride a realm of animal behavior and the world of behavior learned from a society's culture.

A Source for Human Culture

What is the source of culture, society, and sociality or social control? There is little if any evidence that culture arose through biological evolutionary processes. No amount of evidence for strictly biological evolution would establish that human culture "evolved" in a similar manner. In fact the principles leading to sociality and social control are contrary to the principles of random variation and natural selection considered to explain the evolution of biological functions.

Alternatively, culture might have appeared spontaneously, by processes completely independent of any conscious effort or design by human beings. Most societies
seem far too complex for one person to understand or for
a group of people to design.[34] "Some of the rules of law
will be the product of deliberate design, while most of the
rules of morals and custom will be spontaneous
growths."[34] What is the basis of the spontaneous order?
Some who eliminate God rely on super-personal "self-
organizing" forces to create the order, ultimately giving
evolution credit for society's formation.

Few will deny that the acquisition of culture and social
order is the most important step in humanity's progress.
Humanity wants culture to be guided, constrained and not
left to its own desires, with Platonists seeing the direction
provided by something eternal and to the modern-day
positivist by something temporal.[35] For many others it is
best explained in an ultimate sense as part of God's
progressive revelation.

Culture's Value and Truth

Culture is the means by which people are equipped to
realize or fulfill their potentiality, both culture and the
potentiality being essential for that to occur. With human
potentiality but no culture or social control, individuals
remain as the feral child; and without human potentiality
the biological being remains forever no more than an
animal. Normally human beings receive the imprint of
culture and society beginning at birth, determining what
kind of social being they become. But we are not *completely*
determined by that imprint; we are also partly self-made
in the sense that during development especially, but also in
all of life, we can freely choose what habits we will allow
ourselves to form.

If truth is something that must be universal and objective, some beliefs and cultural practices are most important truths. Evidence indicates that some cultural norms are universal, dating from a time so long ago that it is difficult to see how such laws once established in one society could have spread to all others as a means of becoming established universally. They could have been established by revelation of a design (from God, perhaps) to all peoples.

The cultural norm of avoidance of incest, universal throughout human societies, is as old as history, even in societies that have been geographically isolated for ages. Avoidance of incest is not an innate, inherent, physical norm; full brothers and sisters separated from each other from an age so young that they do not remember one another do not have a natural aversion for incest.[36] They express an aversion if they learn that the object of their lust or love is a sibling.

Human beings do not avoid incest because they know that the multiplication and perpetuation of genetic errors is more probable in children of closely related human beings. This knowledge is readily available today but early people did not understand that incest would increase the risk of defective births. To consider avoidance of incest a norm for human beings but not for animals must give this norm value.

A Cultural Norm for Cultural and Biological Progress

Avoidance of incest has been important for human advance both culturally and biologically. Incest avoidance should provide opportunity for variation to cause changes more rapidly in an evolutionary process. Incest or inbreeding maintains and magnifies biological traits, but breeding to a completely different strain promotes the incorporation of new traits. Hybrids produced by crossing different strains are examples of changes that can be rapidly accomplished. Breeding of individuals with very different gene pools will generate new varieties. Any means of accelerating evolution also allows for a faster rate for increasing complexity; it has been suggested that some phases of human evolutionary development have been relatively swift.[37]

Avoidance of incest is also of value because its restrictions permit a greater freedom from the constraints of nature. Humanity has been less subject to the magnification of undesirable genetic expressions, a protection of individuals from harmful genetic mutations and hence enlarging the limits within which their freedom is capable of expression.

This norm is an example of a revelation of a design with no apparent practical human benefit when it was put forth. Some norms came before the formalizing of norms such as by God in the Ten Commandments. Regardless of the cause, avoidance of incest was necesary for progress in the human quest for perfection since people could be victims of their biological processes. The norm was obviously important because of God's command to humanity to "be fruitful and multiply" and "have dominion" over all life on earth.

An Evolutionary Process for Cultural Growth?

A unique feature of the human potentiality permits people to enlarge their freedom within limits through changes in culture rather than by changes in the individual's innate potentiality. Cultural limits for human freedom are illustrated well in the laws given by God to insure that Israel would survive and prosper as a nation designed to carry God's blessing to all people of the world. These Hebrew laws are viewed by some today as narrow, archaic, or even ridiculous. Their purpose was met long ago and the reasons for some of them hardly exist today. The limits of freedom expanded where Christianity replaced the Hebrew religious laws. Tht limits of freedom continue to be enlarged as culture is molded by God's progressive revelation.

The limits of freedom must expand for humans to reach their full potential. On the one hand people cannot live under laws such as those of the old Hebrews, some of which proved impossible to keep; on the other hand our nature is such that we do not want to live by any laws of God or under jurisdiction of any higher authority. The human drive to live without a directing plan or the design of God's laws may appear to be inconsistent with God's interest in humanity. Actually that feature enables people to grow and to realize an ever-greater achievement of their potential so that fewer divine laws are necessary. Growth must include realization of the importance of culture and acceptance of responsibility for an environment in which all people can develop toward the ultimate promise of their potentiality.

Human freedom with constraint is somewhat analogous to the field in which interacting forces constrain the activity of elementary particles at the subatomic level. Without the interacting forces a particle is nothing. Without the interacting forces of culture, individual human beings have no human identity. People by themselves, without the expression of culture's limits on freedom, remain as animals. People may believe they can choose total freedom and become self-made, but they are able to raise themselves up only in the sense that they can freely choose how to develop the potentiality given them.

Where Science and Reason Failed

The means for achieving full expression of the human potentiality have not been found in what science can do to modify culture; they are found only in religious truths. In fact, science cannot be said to have enhanced human development by enlarging the limits of freedom. Some writers say that we have become enslaved to a science-based technology. Humanism proclaims our ability to be self-made, but the power of reason and freedom unconstrained by higher laws have failed to solve our problems. Science has not embellished humanism with anything to make it succeed any better.

The failure of science to answer basic questions about human existence forces one to return to religion for seeking value-laden views of humanity. People may have thought there was a chance to abandon religion's value-laden view of human beings, replacing it with new knowledge from scientific theory, but that was never possible. Before science came into existence, God imprinted humanity with indelible manifestations of the divine revelation; they are stamped into human society and direct the development of human beings. That left us a theological background whether we like it or not and we cannot cast it off or do away with it. God's imprint has been the most impressive and lasting of all aspects of human society.

Human Potentiality: A Creative Design

The potentiality to develop the human uniqueness can be explained as part of a creative design, given to only one species of animal. The design includes plans for sociality which was provided to fulfill a need and plans for science to fulfill another need; both gifts are necessary for achieving a meaningful existence. As the design unfolds, humans are no longer dependent on the survival tactics of the evolutionary processes, and the unique human qualities of empathy and love no longer have to be explained as inborn realizations of the evolutionary descent of humans from animals. Denial of a design with scientific explanations for the potentiality has witnessed a cultural regression from the ideal to varying degrees of the self-interest, self-indulgence, and self-preservation inherent to animals.

The Bible states that God breathed life into human beings, creating the unique mental processes that endow people with the intellectual powers for dealing with the unperceived, imperceptible, and unimaginable. The uniqueness is a potentiality totally dependent on a cultural imprint for its full development. The potentiality is a freedom with which humans can grow and approach the realization for which they were designed.

Chapter 10

The Purpose of Science

Science as knowledge is one of humanity's most important possessions. Its purpose and timing have been crucial for insuring human survival. It gave us prediction and control at the right time to meet some vital human needs. As we have noted, its purpose is not the development of scientific theory to answer ultimate human questions. And we must acknowledge that the impressive steps made in the last century to predict and control our environment actually owe more to technology than to science.[1]

A Plan for Survival

Without scientific knowledge to insure their survival, earlier peoples worked out systems of morals and ethics for their wellbeing. Examples are revelations of God to the early Hebrews and the philosophy of the early Greeks. Besides being used to develop new and improved patterns for survival, those two systems were blended in molding the life, death, and resurrection of Jesus Christ into the theology of Christianity. (They were also used in developing Islamic theology.) A well-developed theology was necessary for Christianity's maturation and for its indelible imprint on Western society.

The contribution of early Greek philosophy culminated in Thomas Aquinas's amalgamation of Aristotle's philosophy with Christianity. Earlier scholars such as Justin

Martyr, Clement of Alexandria, Origen, and Augustine had used Plato's teachings in a similar way to develop the theology of the early Christian church.[2] Early Greek philosophy appeared at an appropriate time and with a content that met the needs for fulfilling the human potentiality. Thus, the blending of concepts in Greek philosophy and the revelation of God to the Hebrews was well suited to bring Christianity to maturity. That happened in the interlude immediately before and during the Middle Ages, the "dark ages," when science paused but not humanity, which developed Christian theology. Why such a contrast in achievement in various forms of human thinking?

A Pause with Purpose

One explanation for the contrast is that the period before and during the Middle Ages was being used by God for maturation of the human creation before a need for science had arrived. Humanity needed both a philosophical and theological basis to prepare for the development of science. Had science come earlier and revealed an empty and meaningless universe (as it seems to have done recently) people might not have been able to survive. Recent scientific speculations have been less destructive because God has already revealed the design for humanity through the theological development of Christianity. Human beings would hardly have survived with science alone; science would not have enabled us (as the humanist suggests) to work out all our questions and problems.

The human potential for thinking and reasoning was to be developed before we would receive the gifts of science and technology. That preparation provided people with sources of value for their existence, a prerequisite for receiving science. The content of the prerequisite may have been presented as "complete." Alfred North Whitehead commented that "the safest general characterization of the European philosophical tradition is that it consists in a series of footnotes to Plato."[3] Mortimer Adler has written that little substantial has been added to philosophy in twenty-five centuries.[4] The source for Christian theology's content has been sealed and completed for centuries. Thus it could be said that humanity has not received anything substantially new.

Technology, or what might be called "practical science," eventually became essential to human survival. Technology was less important to survival in earlier times; hence the gap of more than fifteen centuries between the laying of the scientific foundations and the construction of the edifice for prediction and control. Only when there was an absolute requirement for survival as a species were human beings presented with science as the means for technology's implementation.

Science seems to be inherent in a design to insure our survival as a species. When people were ready for it, science for developing prediction and control appeared and when people needed it, practical science also appeared. An absolute need did not appear until tools for moral behavior had been developed. Without the establishment of moral behavior during the infancy of their species, human beings might not have survived, or might have tried to reason their way toward moral behavior. Without such cultural roots imprinted much earlier, the new philosophy of science could be destructive to human existence. The unfolding and implementation of the design for insuring survival gives the universe purpose and meaning.

Neither Determinism or Indeterminism

Is every detail of the design predetermined? If total supervision and direction were designed for all activity in the universe, everything would be determined. Advocates of that concept, *determinists*, deny that there is freedom of choice or that anything could happen by chance since the designer or God is the cause or mover of everything. At the other extreme, *indeterminists* propose that everything results from no cause and occurs simply by chance. Historically, Western society believed that their existence was completely determined by God until science opened a door to indeterminacy in the past four to five centuries. But a rigid determinism was not there in "the beginning."

The biblical creation story is inconsistent with the absolutes of either determinism or indeterminism. The first chapter of Genesis describes an order, organization, and set of laws governing the universe, reflecting a God-determined universe that operates to control every minute detail. God knows and directs all things during all ages into the past and into the infinite future. But the second chapter of the Genesis creation narrative describes a freedom of choice, introducing the possibility of events being based on chance. Adam is able to choose fruit from all but one tree in the Garden of Eden. Freedom is also shown in the prerogative God gives Adam to name every living creature.

This blend of determinism with indeterminism indicates that God did not intend to dictate every detail of every activity in the universe. The freedom given us in a world with prescribed laws and detailed organization is a *freedom within limits*. When the theologians' absolute form of determinism was discarded, it was replaced by the scientist's indeterminism, which was nearly as absolute. An unanswered question is how much of our universe should be attributed to determinate manifestations of a design and how much to indeterminate signs of freedom of choice or chance.

Modern scientists, like the early Greeks, maintain at least a trace of determinism because they speak of universal laws or a cosmic code, symmetry, beauty, order, harmony, and the possibility of mathematical explanation, even in a universe they consider at the worst indifferent, meaningless, and uncaring for human life. Belief in a mathematically describable universe promotes a desire to understand the universe as well as possible, with the hope that such understanding might broaden and deepen our insights. This scientific faith persists despite evidence from the physics of our century that such intimate knowledge is unattainable.

Humans Seek Comfort in Order, Beauty, and Knowledge

Centuries ago deists believed that God created the universe but assumed an indifference toward the creation thereafter. Deistic scientists attributing a law, code, mathematics, or order to the design and operation of the universe, at the same time viewed the universe as meaningless and indifferent to human beings. Many modern scientists are deists; the object of their faith is found in unifying concepts, order, symmetry, life on other planets, and a mathematical basis for the universe.

Some scientists completely deny any universal "law and order" underlying the universe; they believe that all life developed merely by chance and that life's only purpose is to sustain itself. We are alone in the universe, they say, with human beings as the only source of value and purpose and with science as our only source of truth.[5] Why should we value science as our only source of real knowledge, meaning, and moral decision if everything happens merely by chance? For people who find value and purpose only in themselves, how can they say we should rise above ourselves, to find transcendence?[5] For that, we need another source of real knowledge; transcendence comes not by reason or scientific method. Something transcendent is beyond complete understanding; it is outside the realm of human experiences perceived by the five senses.

In an indeterminate, meaningless universe where every event is based on chance, neither science nor people trying to raise themselves up to the level of transcendence will find answers to human aspirations or an understanding of their significance. Neither science nor humanism provides satisfying answers.

From Unsatisfied Expectations to Great Uncertainty

Modern physics, providing its explanations and predictions in probability terms, has been heralded as a kind of ultimate knowledge. The idea that theoretical physics is actually philosophy was expressed by the physicist Max Born and is undoubtedly shared by many other scientists.[6] But that science falls short of discovering the knowledge people seek. Science in general has failed to satisfy human expectations; prediction and instrumental control are not enough. People want what Adam sought by eating fruit from the tree of knowledge; understanding based on probabilities is not sufficient.

Science as Philosophy?

The evidence for Darwinian evolution is no greater than that for quarks or black holes, making all of them as severely underdetermined as the metaphysical concepts of God and the cosmos. That situation does not deny validity for some scientific theories of evolution, but it poses a difficulty in selecting one theory over others.

Metaphysical concepts of God and the universe have been rejected because they lack validation based on scientific methods of observation and reasoning. Positivistic science demands that any concept be verified that way.[7]

When physicists become philosophers, however, they abandon the rigorous scientific method necessary to validate their theoretical constructions. Quarks, mesons, and black holes have a very tenuous empirical basis, but they now dictate answers to philosophical questions.

Evolution as such does not conflict with concepts of God and a design for the universe. Considering such attributes of God as omniscience, omnipotence, and omnipresence, one cannot deny that God could have used evolutionary processes as the means to create human beings. But dogmatic evolutionists demand acceptance of all details of Darwinism; anything less than total commitment to random variation and natural selection would open the door to determinism, allowing some purpose and direction in evolutionary processes. There will never be enough empirical evidence to prove it either way, however.

As discussed earlier, the extreme improbability of self-assembly of twenty to thirty amino acids into not billions of different possible proteins but into the relatively few required for life to begin means that the evolutionary process needs a direction and a meaning or purpose. And, since the four physical forces of the universe have exactly the right properties to produce a universe supporting life, there is another important reason that evolution proceeded one way and only one way. If there was direction in creation of the four physical forces, then all that follows must have direction and some meaning. Conversely, if chance is the basis of all events in the evolutionary process, all that follows must be totally indeterminate, meaningless, and uncertain; no value can be placed on anything; it makes no difference what chance might cause to happen; people become merely a higher form of animal.

Science Is Not a Source of Value

The most profound human values were not established or embellished by science; they were cherished by people long before science appeared. While science and technology have shown great accomplishments, many cultural abilities have been lost: the ability to maintain stable family structure and interpersonal relationships; to insure survival from inadequate nutrition and from crime against individuals and groups; and to maintain a belief in any worth or meaning for the human condition. The traditional social value of wanting to better humanity has not been in any part due to science—other than in the material prosperity enjoyed by a minority of the world's population. That represents a conflict between science and the human condition based on profound human values.

Science's Purpose: A Value that Must be Science-free

Belief that science has a purpose is as valid as the scientist's view that the universe is comprehensible. If science's purpose is to predict and control, that purpose has been realized, at least to some extent. That gives science value, but scientists argue that science must be value-free or value-neutral. A value-neutral human pursuit is one followed with the provision that no value, one way or another, will be held for the pursuit or its outcome. Thus the value of people or any interests they might have must be disregarded. Scientists seeking understanding and wisdom from their work do place a value on science, however.

Conversely, one could say that value must be science-free, meaning that scientists using factual knowledge from scientific endeavors are not best qualified to theorize on such topics as cosmology and evolution. Trying to do so has generated "wild" ideas based primarily on such values as desire for order, symmetry, beauty, and comprehensibility, but backed up with little evidence.

That science has a purpose is a value that should be science-free. Science is essential for satisfying human needs; its ultimate goal is to guarantee human survival. That value is not dependent on any knowledge of physics or evolutionary processes. More important, value must be science-free to protect us from the consequences of postulating a meaningless existence in a universe where the most fundamental processes are assumed to operate in a random way, with all events being a matter of chance occurrence. The assertions that science is a gift, given to human beings for a reason, and that it has purpose and value, do not violate the scientist's condition that science must be value-free in its methods, nor does it violate the stipulation that values must be science-free.

Science Insures God's Promise

Religion, as the only source of humanity's value-laden ideas, should be the means for the most important expression of society. Theology should not be concerned with conflicting ideology of current cosmology, evolution, or reductionist biology, because the scientific theories growing out of these will not survive. They will be rejected not by religion but from within, as happens to all value-neutral myths that masquerade as theoretical truths but are actually severely underdetermined. Theology needs only to vigorously assert the truth where it alone provides the source: in meaning and value-judgment for human beliefs.

Theology was not intended by God for establishing empirical facts; therefore it has no empirical truth value. Theoretical science being so underdetermined cannot have truth value, and theology cannot conflict with something else that has no truth value. Theology expresses the importance of nature-to-human and human-to-nature relationships, placing a value on people as part of the universe. Of course such claims are theoretical or ideological and are in general as severely underdetermined as those of scientific theory.

The Covenant Is Implemented through Science

Theology claims that human beings will survive as a form of life. The design of God includes a promise found in the Christian and Hebrew Bible protecting the human race from being wiped off the face of the earth.[8] That promise contains a provision giving human beings a capacity to recognize and manage the problems encountered with their physical existence. God's provision to human beings is an understanding of science for dealing with those problems; the timing of the provision is appropriate to the appearance of the need. The question of what purpose science serves is answered by an examination of the limits of science. Science shows how human populations could be decimated by an interaction between overpopulation and disease. It can be argued that there is one truth, which is both objective and universal, in the idea that God gave human beings science for insuring their survival and not for gaining understanding or wisdom; the limits of possibility are defined. That places a value on science, and theology could maintain it to be science's only purpose. God gave people science as part of an ongoing revelation to insure human survival.

A Progressive Revelation

The process as well as the purpose of the world's creation and subsequent evolution is a progressive revelation of God's design. Revelations have come through the Bible and through tools needed for survival. Recent revelations of scientific knowledge are intended to insure survival in a world with increasing human population.

Progressive revelations are evolutionary in following God's plan; like other evolutionary processes they are programmed according to a schedule. Human beings have been frustrated by their inability to alter that timetable by increasing scientific activity. The time appropriate for people to gain understanding is determined by the design, not by human desires. Scientific progress is not directly proportional to resources allocated to research. Many great scientific achievements were realized with few supporting resources. With a design unfolding in an evolutionary process, creative input to empower the imaginative-intuitive-insight process determines the appearance of important scientific achievements.

The progressive revelation has the potential for enlarging the limits constraining human freedom. The limits are enlarged by growth in understanding. Such growth is in understanding for both the human subject as interpreter and the rest of the world as objects, as they mutually determine one another.[9] This relationship is evident for a person as an interpreter and the text of the Holy Bible as an object. For many that text never ceases to be a source of new and progressive understandings.

That text is not finished but is continuously "determined" by people who find new meanings in a content that is virtually unchanged for more than nineteen centuries. Rethinking on an old text becomes a means for progressive revelation, a way that people can increase the limits of their freedom. With no understanding the text appears to be merely a series of objective formalized laws for restricting any growth. With understanding the text speaks to individuals and by "determining" that person, progressive revelation makes human growth possible.

Chapter 11

In the Beginning There Was Value

Can Value Come from Human Reason?

As we have said, the basic questions are who we are, why we are here, and what is our destiny in a life and world that appear meaningless. Our answers to these questions must insist on value being assigned to human beings and their universe. Value-neutral scientific theoretical explanations are unacceptable; their description of a meaningless existence in a value-neutral universe is based on human frustrations in being unable to gain understanding from the "tree of knowledge." Science has not given human beings value because it has been unable to give them understanding. Using only their powers of reasoning, people are unable to determine what is of ultimate value.

When human reason is assumed to be the final determinant of the value for anything, values will be (a) relative to the society's wants and desires, (b) without the objectivity and universalism to allow them to approach truth, and (c) equal in the sense that none can be considered better than the next. In reality, nothing will be of value. Human reason is not necessary to define the value of meeting the needs of all human beings.

The realization of everyone's *needs* is an objective and universal value, representing a single natural right from birth. All other human values are based on people realizing their desires and wants. Because all scientific truth must be based on value-neutral and objective criteria, *science* has no basis (or value) for meeting all people's needs. That is, there is no more value to *that* goal than to anything else people do. Some individuals would deny science any role for value determination and would attempt to establish values that are relative and subjective, but such values can claim no universal and objective truth.

With random variation and chance as the basis for all events, nothing happens for any rational purpose; there is an irrationality to all aspects of the universe. In contrast, an objective, universal rationality expects order and value to characterize what happens in the universe. There can be no rationality without value; with irrationality comes loss of value. If human life is viewed as meaningless in an indifferent universe, it seems consistent for irrationality to be apparent everywhere in life. Such irrationality would hardly aid humanity in solving its social problems.

Many people have attributed human irrationality to primitive and constraining religious, moral, political and economic creeds coming out of the Dark Ages; the age of reason and science was expected to change that. Some people would argue that irrationality persists now only because the creeds have remained unchanged—they have been fiercely protected against change—and that such creeds produce an irrationality in people who continue to believe them.[1] But the belief persists that reason is capable of equipping us to solve our problems; since it has not yet been seriously applied only intermittent progress has been made in realizing value in human life.[2]

The academic enterprise is sometimes blamed for the persistence of a damaging irrationality.[3] Has it failed to develop a kind of organized inquiry wholly devoted to the promotion of rationality? That is exactly what academic inquiry does; it uses human reason to seek what is rational in life, with the proviso that truth must be objective and universal. The academic enterprise has denied a role for the subjective but has failed to solve human problems with truths sought in the objective and universal. To solve basic human problems academic inquiry would have to go beyond human reason. Constructive action occurs only when a creed imprinted in culture through religion becomes as important as inquiry by reason and science.

God Gave Value with Human Life

The secular academic enterprise cannot establish values for students as long as it believes that scientific method alone is the way to truth. Religiously based academic enterprises *can* find values, and relate them to the knowledge found through reason. We must have both values and reason for solving humanity's problems; that is no value-neutral enterprise. To solve our problems without need of any values from our religious cultural heritage, we would have to be innately good. What is there to motivate us to consider another person's problems as our own?

Although irrationality can certainly be found in religious creeds maintaining extreme positions, God's revelations provide the only effective rationality for solving human problems. Religion succeeds where social inquiry fails to promote understanding between people and where natural science fails to promote human cooperation in using and sharing the world's resources.

Social science has introduced new ways to perpetuate irrationality, by teaching people more effective means of manipulating others.[4] Manipulation in technologically advanced societies is evident in the methods politicians use to get elected and in the advertisements industry uses to sell its products. The social sciences have not been effective in promoting human progress; any success they have achieved has required people to be obedient and incapable of innovative thought and action.

People try to establish objective, universal values themselves but fail in the attempt. They rather adopt values meaningful only to their direct interests; such values are not objective and universal. Religion has been the only means of establishing objective and universal values; that cannot be achieved by a cooperative effort of human reason.

People must continue to allow religion to decide what is of value. With value so decided, people will place less value on what they desire, and value will approach objective and universal truth. Adopting this position does not force people to sacrifice any intellectual independence or individual freedom, but does keep people from regarding those two assets as supreme.

A humanist might argue that we should not surrender our intellectual independence. By continuing to use our reason, intelligence, and experience, they say, we will discover and help create what is of value. The ability of human rationality to create truth about what is of value may be no more than a illusion, however; it is evident that much of what is of value has come into existence unforeseen and unintended, evolving gradually in time. The creation or appearance of gradually evolving values, in past centuries through religion (where God is the creator of values), did not cause people to sacrifice their individual freedoms. On the contrary, it increased human freedom.

Obedience and Freedom: A Tension for Human Growth

Those who regard the Genesis creation story as a myth may not yet have realized that the "wages of self" is the way of death. Only knowledge from God leads to understanding and wisdom; knowledge from self is merely knowledge, not representing truth. The few biblical figures who "walked with God" did so by seeking and following his truth rather than by following their own knowledge and doing "what was right in their own eyes."

The Old Testament laws were instituted primarily to insure survival of the Hebrew tribe as a people and to safeguard the promise that they would become a blessing to all the world. The covenant supporting those promises could not survive where people went the way of self-direction. A covenant of obedience is essential for people to grow into the fullest expression of their human potential. Without such a covenant we have difficulty rising above the mere animal level to realize the unique gift given us by God. The covenant also implies that we cannot achieve innocence, happiness, and a childlike state. God has given humanity a potential to grow out of childhood and develop into beings with whom he can have a meaningful and lasting relationship.

The depth of the biblical story is unique in its portrayal (a) of the relationship extended by a concerned God, the creator, to his human creatures, and (b) of the tension that develops between God's plans and the actions of free-willed humanity. In many contemporaneous cultures the gods had little compassion for human beings after the creation. Considering the fact that many attribute the Genesis account to ancient contemporaneous mythology, it is interesting to note that, in Genesis, God's first opponent describes the divine plan as a myth. God's adversary, in the form of the tempter, tells the first woman, "you will not die"; that is, what God said is a myth. To regard the Genesis episode as a myth may free a person from obligation to live under God's plan, but it does not erase the anguish of a life that is not eternal.

By our divinely given free will, we continue to choose for ourselves whom to serve. The book of Judges records that "In those days there was no king in Israel, every man did what was right in his own eyes."[5] In modern language:

> It is ethically right to aim at whatever will promote the increasingly full realization of those values which are more intrinsically or more permanentingly satisfying, or involve a greater degree of perfection.[6]

The laws of God handed down by Moses and Joshua were mandates, but God's people followed them only when they found it convenient to do so. Ancient peoples had accumulated little knowledge to justify doing "what was right in their own eyes." That which is right is based on truths. The truths for ancient people to live by were given as part of God's revelations.

People Seek Value and Find Freedom

With people's growth and maturation in their relationship with God, the values established by the creator matured from the constraining laws of the Old Testament to a new kind of freedom that people could choose, expressed in the theology of the New Testament. People internalize values and gain wider freedom as they grow in understanding God's revelations. Within limits, freedom is enlarged by God's continuing revelation and by realization of universal and objective values by human beings.

Without methods for determining what is of value (including our feelings and the content of religious books and institutions) doubt about the meaning and value of our lives is inevitable. The person who knows God looks there for meaning and value; for those obedient to God's design, that represents an infallible method. The person who believes that there are no infallible methods and denies God a role in establishing value must hope that people can spin out their own values.

Progressive Evolutionary Revelation of Value

The closeness of human beings to God increases as they make progress in an evolutionary process of revelation. God directs individual growth so that people develop into the persons God intended them to be. Those individuals become more intimate with God and human survival also becomes more important to God. Humanity becomes better protected from the natural forces operating to limit the population of all living matter.

In the Beginning There Was Value

In the beginning, human beings gained a value for their existence through God's revelation of a design and a means for salvation. People lost that original value as they began to believe that science, technology, and education would promote human welfare and show humanity what is genuinely of value in life. Some thinkers believe that:

> The evolutionary biologist has at last undercut this "old covenant," linking nature and value; and has made it necessary for scientists to construct a "new covenant," in which science will take over from philosophy and teach men to live by values free of any cosmic sanction.[7]

So humans seek to regain value by building rules of rational problem-solving, hoping to settle on what is of value in order to enhance their freedom, creativity, capacity to love, and wisdom. By rationally employing those rules, some people hope to interrelate objective facts, desires, experience, feelings, and insights in order to determine what is of value in the world. A specific aim of science is "to improve knowledge of humanly valuable truth."[8] The dilemma faced by the rational mind is that so much input is subjective, approaching even the metaphysical, which for that kind of mind is no basis for truth.

To the intellect admitting only objective, universal knowledge as a basis for truth, what is of value can exist only in abstract form with no real basis at all; personal feelings cannot be considered. The potentially corrupting influence of what scientists consider to be largely irrational subjectivity is not allowed by them to interfere with and ruin the designs of reason and science. But the objective, universal, and value-neutral truths of science can never promote human welfare and show people what is of value.

In the Garden of Eden, the first human beings did not understand the value of human life or the value of anything else; Adam and Eve were little different from most people today. Like modern humans, they heeded no value system, making them seek knowledge as the means to enhance their existence. Like Adam and Eve, people today seek knowledge but not to determine what is of value in life; they have already decided that for themselves. There can be no objective, universal truth because there will never be a consensus on what is of value; value is determined by what people desire for themselves, even if of doubtful value to others.

Does science have the potential for satisfying human yearnings? According to Aristotle:

> Our highest activity and hence happiness is to be found in the pursuit of scientific knowledge—the knowledge of the cause of things—or in short, a knowledge of ultimate reality.[9]

In contrast stands the revelation of God long before Aristotle:

> But where shall wisdom be found? And where is the place of understanding? Man does not know the way to it, and it

is not found in the land of the living. The deep says, "It is not in me", and the sea says, "It is not with me" . . ." It is hid from the eyes of all living, . . . God understands the way to it and he knows its place.[10]

But is there a way for human beings to use science to study the natural world and thereby find the meanings God has for them? About finding wisdom in God's natural world, the Bible says "Lo, these are but the outskirts of his ways; and how small a whisper do we hear of him."[11] The handiwork of God is indeed seen in the world of nature. To find God's purpose, however, we must look to the revelation designed for that.

People of God know who they are; they have meaning and purpose for their lives; they know their destinies. Over three thousand years ago Joshua asked his people to "choose this day whom you will serve . . . ."[12] Today, science is chosen by many as that which they will serve. Ironically, science and reason have tried to show people that their existence is without meaning. For many scientists, the pleasure of pursuing science is fulfilling, but science may actually make life more meaningless, except for providing some material improvements. All the beauty, order, symmetry, and balance of the natural world, no matter how pleasing, do not add up to the understanding we seek for ourselves in the cosmos.

God gave us science as a part of the revelation of all creation's purpose and design. God gave us science to insure our survival, not to "explain" ourselves. Following the Hebrew psalmist we can still have faith that "He who dwells in the shelter of the Most High, who abides under the shadow of the Almighty, will say to the Lord, 'My refuge and my fortress; my God, in whom I trust.'"[13] For that faith to be assured, we too must heed the words of Joshua: "Choose this day whom you will serve . . . ." The gods our fathers have served may have been human reason and science, "but as for me and my house, we will serve the Lord."[14] When belief helps us discover who God is, we begin to learn who *we* are, why we are here and what the world really is.

Bibliography

In the Beginning

1. Robert Davidson, The Cambridge Bible Commentary, Genesis 1-11 (London; Cambridge University Press, 1973), p. 13.

2. Gareth Nelson and Norman Platnick, "Systematics and Evolution", in Beyond Neo-Darwinism, ed. MaeWan Ho and Peter T. Saunders (New York: Academic Press, 1984), p. 145.

3. Will Durant, The Story of Civilization: Part II The Life of Greece (New York: Simon and Schuster, 1939), p. 139, 356-357.

4. Mary Hesse, Revolutions and Reconstructions in the Philosophy of Science (Bloomington: Indiana University Press, 1980), p. 168.

5. The Holy Bible. Revised Standard Version, Gen 2:7

6. Ibid., Gen 2:17.

7. Davidson, p. 34.

8. Ibid., p. 34-35.

9. Ibid., p. 35 (see Deut. 18:14).

10. Diogenes Allen, Philosophy for Understanding Theology (Atlanta: John Knox Press, 1985), p. 34.

11. Davidson, p. 29, 34.

12. Nahum M. Sarna, Understanding Genesis (New York: McGraw-Hill, 1966), p. 26-27.

13. The Holy Bible, Gen 3:4.

14. Ibid., Gen 3:5.

15. Amos Kidder Fiske, The Myths of Israel (London: MacMillan Co., 1897), p. 57.

16. F. A. Hayek, Law, Legislation and Liberty (Chicago; University of Chicago Press, 1973), p. 10.

17. J. G. Crowther, A Short History of Science (London: Methuen Education, 1969), p. 4.

18. Leon R. Kass, Toward a More Natural Science (New York: The Free Press, 1985), p. 131.

19. J. Bronowski, The Ascent of Man (Boston: Little, Brown & Co., 1973), p. 437-438.

Science in the Beginning

1. J. G. Crowther, A Short History of Science (London: Methuen Educational, 1969), p. 4.

2. Colin A. Ronan, Science: Its History and Development Among the World's Cultures (New York: Facts on File, 1982), p. 62.

3. Will Durant, The Story of Civilization: part II The Life of Greece (New York: Simon and Schuster, 1939), p. 137.

4. Stephen Toulmin, The Return to Cosmology (Berkeley: University of California Press, 1985), p. 223.

5. Vern L. Bullough, ed. The Scientific Revolution (New York: Holt, Rinehart and Winston, 1970), p. 2.

6. Bertrand Russell, Wisdom of the West (Garden City, N.Y.: Doubleday & Co., 1959), p. 44-45.

7. Durant, p. 144.

8. Russell, p. 101.

9. Ibid., p. 100.

10. Ibid., P. 101.

11. Durant, p. 635.

12. Ronan., p. 72.

13. Russell, p. 173.

14. Crowther, p. 14.

15. Ibid., p. 22.

16. Russell, p. 81.

17. Durant, p. 635-636.

18. Ronan, p. 118.

19. Ibid., p. 121.

20. Ibid., p. 103.

21. Russell, p. 186.

22. Diogenes Allen, Philosophy for Understanding Theology (Atlanta: John Knox Press, 1985), p. 118.

23. Durant, p. 527.

24. Ibid., p. 528.

25. Ibid., p. 531.

26. Ronan, p. 108.

27. Ibid., p. 103.

28. Russell, p. 90.

29. Edward Hussey, Aristotle's Physics Books III and IV (Oxford: Clarendon Press, 1983), p. 138-175.

30. Ibid., p. 168.

31. John Maddox, "How Special is Special Relativity?", Nature, 313 (7 February 1985), p. 429.

32. Heinz R. Pagels, The Cosmic Code (Toronto: Bantam Books, 1982), p. 36.

33. Ibid., p. 122.

34. Hussey, p. 188.

35. Pagels, p. 56.

36. Ibid., p. 243.

37. Hussey, p. 122.

38. Ibid., p. 130.

39. Heinz R. Pagels, Perfect Symmetry (New York: Simon and Schuster, 1985), p. 186ff.

40. Hussey, p. XXXV.

41. Durant, p. 633-634.

A Renewal of Interest in Science

1. T. S. Kuhn, The Structure of Scientific Revolutions, (Chicago: Chicago University Press, 1962).

2. Vern L. Bullough, ed. The Scientific Revolution (New York: Holt, Rinehart and Winston, 1970) p. 2.

3. Diogenes Allen, Philosophy for Understanding Theology (Atlanta: John Knox Press, 1985), p. 17.

4. Gabriel Moran, Theology of Revelation (New York: Seabury Press, 1979), p. 52.

5. Bullough, p. 4.

6. Ibid., p. 4-5.

7. Leon K. Knoebel, "Secretion and Action of Digestive Juices," in Physiology, 3rd ed., ed. Ewald E. Selkurt, (Boston: Little, Brown & Co., 1971), p. 615.

8. Edwin A. Burtt, "Mathematics, Platonism, and the Renaissance," in The Scientific Revolution, ed. Vern L. Bullough (New York: Holt, Rinehart and Winston, 1970), p. 31.

9. Will Durant, The Story of Civilization: part V, The Renaissance (New York: Simon and Schuster, 1953), p. 528.

10. Giorgio de Santillana, "The Crime of Galileo", in The Scientific Revolution, ed. Vern L. Bullough (New York: Holt, Rinehart and Winston, 1970), p. 28.

11. Bertrand Russell, Wisdom of the West (Garden City, N.Y.: Doubleday & Co., 1959), p. 184-185.

12. Heinz R. Pagels, The Cosmic Code (Toronto: Bantam Books, 1982), p. 307.

13. Ernst Cassirer, "Astrology and the Development of Science", in The Scientific Revolution, ed. Vern L. Bullough (New York: Holt, Rinehart and Winston, 1970), p. 47.

14. Colin A. Ronan, Science: Its History and Development Among the World's Cultures (New York: Facts on File, 1982), p. 203.

15. Ernst Mayr, The Growth of Biological Thought (Cambridge, Mass: Belknap Press, 1982), p. 84.

16. Ronan, p. 245-246.

17. John Marks, Science and the Making of the Modern World (London: Exeter, 1983), p. 238.

18. J. G. Crowther, A Short History of Science (London: Methuen Educational, 1969), p. 22.

19. Ronan, p. 253.

20. Hans Baron, "Humanism and Science: I", in The Scientific Revolution, ed. Vern L. Bullough (New York: Holt, Rinehart and Winston, 1970), p. 41.

21. Harcourt Brown, "Did Science Depend Upon Renaissance Concepts?", in The Scientific Revolution, ed. Vern L. Bullough (New York: Holt, Rinehart and Winston, 1970), p. 56.

22. Ibid., citing Butterfield, p. 60.

23. Ronan, p. 186.

24. Marks, p. 228.

25. Ibid., p. 237.

26. Ronan, p. 186.

27. Walter J. Ong, "Printing and Science", in The Scientific Revolution, ed. Vern L. Bullough (New York: Holt, Rinehart and Winston, 1970), p. 77.

28. Ibid., p. 77-78.

29. Alexander Keller, "Has Science Created Technology?", Minerva, 22 (Summer 1984), p. 160-182.

30. Ronan, p. 256.

31. Keller, p. 172, quoting F. R. Jevons.

32. A. C. Crombie, "The Continuity of Scientific Developments", in The Scientific Revolution, ed. Vern L. Bullough (New York: Holt, Rinehart and Winston, 1970), p. 102.

33. Crombie, p. 102.

34. Mayr, p. 29.

35. Crombie, p. 102ff.

36. Mayr, p. 87.

37. Crombie, p. 102-104.

38. Ibid., p. 103.

39. Peter Brian Medawar, Induction and Intuition in Scientific Thought (Philadelphia: American Philosophical Society, 1969), p. v.

40. Crombie, p. 106.

41. J. G. Crowther, A Short History of Science (London: Methuen Educational, 1969), p. 91.

42. Ernest A. Moody, "Galileo and His Precursors", in The Scientific Revolution, ed. Vern L. Bullough (New York: Holt, Rinehart and Winston, 1970), p. 108ff.

43. Thomas S. Kuhn, "The Causes of the Scientific Revolution", in The Scientific Revolution, ed. Vern L. Bullough (New York: Holt, Rinehart and Winston, 1970), p. 123.

44. Joan Gadol, "Humanism, Natural Science, and Art," in The Scientific Revolution, ed. Vern L. Bullough (New York: Holt, Rinehart and Winston, 1970), p. 64.

Rethinking

1. Peter Brian Medawar, Induction and Intuition in Scientific Thought (Philadelphia: American Philosophical Society, 1969), p. 14ff.

2. A. C. Crombie, "The Continuity of Scientific Developments," in The Scientific Revolution, ed. Vern L. Bullough (New York: Holt, Rinehart and Winston, 1970), p. 107.

3. Medawar, p. 24ff.

4. Ibid., p. 31.

5. Friedrich Nietzsche, The Will to Power, trans. by Walter Kaufman and R. J. Hollingdale (New York: Random House, 1967).

6. Medawar, p. 15.

7. Immanuel Kant, The Critique of Pure Reason, trans. by Norman Kemp Smith (New York: Modern Library, 1958).

8. Medawar, p. 31.

9. Ibid., p. 1.

10. Ibid., p. 55.

11. Richard Rorty, "Pragmatism and Philosophy," in After Philosophy, ed. Kenneth Baynes, James Bohman, and Thomas McCarthy (Cambridge, Mass: The MIT Press, 1987), p. 46.

Rethinking Human Origins

1. Sidney W. Fox, "Protenoid Experiments and Evolutionary Theory", in Beyond Neo-Darwinism, ed. Mae-Wan Ho and Peter T. Saunders (New York: Academic Press, 1984), P. 47.

2. G. Ledyard Stebbins and Francisco J. Ayala, "The Evolution of Darwinism", Scientific American, 253 (July 1985), p. 72-78.

3. Fox, p. 19.

4. Ibid., p. 47. 5. Ibid., p. 52.

6. Koichiro Matsuno, "Open systems and the Origin of Photoreproductive units", in Beyond Neo-Darwinism, ed. Mae-Wan Ho and Peter T. Saunders (New York: Academic Press, 1984), p. 82-83.

7. Ibid., p. 83.

8. Jeffrey S. Wicken, "On the Increase in Complexity in Evolution", in Beyond Neo-Darwinism, ed. Mae-Wan Ho and Peter T. Saunders (New York: Academic Press, 1984), p. 89.

9. Ernst Mayr, The Growth of Biological Thought, (Cambridge, Mass.: Belknap Press, 1982), p. 690.

10. Wicken, p. 93.

11. Will Durant, p. 353.

12. Wicken, p. 109.

13. Ibid., p. 106.

14. Heinz R. Pagels, The Cosmic Code (Toronto: Bantam Books, 1982), p. 112.

15. Mayr, p. 510-525., 566-570., Stebbins and Ayala, p. 82.

16. G. Ledyard Stebbins, "Genetics Prof develops Theory of Evolution", The Davis Enterprise, May 31, 1985, p. 10. also see Mayr, p. 617.

17. Mayr, p. 793-794.

18. Ibid., p. 626.

19. Ibid., p. 627.

20. Ibid.

21. Ibid., p. 625.

22. Ibid., p. 42-43.

23. Ibid., p. 38, 87.

24. Ibid., p. 505.

25. Ibid., p. 209ff.

26. Gareth Nelson and Norman Platnick, "Systematics and Evolution", in Beyond Neo-Darwinism, ed. Mae-Wan Ho and Peter T. Saunders (New York: Academic Press, 1984), p. 145.

27. Elisabeth S. Vrba, "Patterns in the Fossil Record and Evolutionary Processes", Beyond Neo-Darwinism, ed. Mae-Wan Ho and Peter T. Saunders (New York: Academic Press, 1984), p. 115.

28. Gerry Webster, "The Relations of Natural Forms", in Beyond Neo-Darwinism, ed. Mae-Wan Ho and Peter T. Saunders (New York: Academic Press, 1984), p. 193.

29. Vrba, p. 138.

30. Nelson and Platnick, 144.

31. Brian C. Goodwin, "A Relational or Field Theory of Reproduction and its Evolutionary Implications", in Beyond Neo-Darwinism, ed. Mae-Wan Ho and Peter T. Saunders (New York: Academic Press, 1984), p. 224-228.

32. Mae-Wan Ho, "Environment and Heredity in Development and Evolution," in Beyond Neo-Darwinism, ed. Mae-Wan Ho and Peter T. Saunders (New York: Academic Press, 1984), p. 285.

33. Miranda Robertson, "The Proper Study of Mankind," Nature, 322 (3 July 1986), p. 11.

34. Goodwin, p. 219.

35. Ibid., p. 221.

36. Mayr, p. 598.

37. Goodwin, p. 229.

38. Ibid., p. 230.

39. Mae-Wan Ho, p. 268.

40. Ibid., p. 285.

41. Ibid.

42. Jeffrey W. Pollard, "Is Weismann's Barrier Absolute?", in Beyond Neo-Darwinism, ed. Mae-Wan Ho and Peter T. Saunders (New York: Academic Press, 1984), p. 293ff.

43. Ibid., p. 293.

44. Ibid., p. 303.

45. Mayr, p. 622.

46. Wu Rukang and Lin Shenlong, "Peking Man", Scientific American, 248 (June 1983), p. 89.

47. Chris Sinha, "A Socio-Naturalistic Approach to Human Development", in Beyond Neo-Darwinism, ed. Mae-Wan Ho and Peter T. Saunders (New York: Academic Press, 1984), p. 331ff.

48. Ibid., p. 334.

49. Ibid., p. 337.

50. Werner Stark, The Social Bond vol I (New York: Fordham University Press, 1978), p. 105ff.

51. Hans Blumenberg, "An Anthropological Approach to the Contemporary Significance of Rhetoric," in After Philosophy, ed. Kenneth Baynes, James Bohman, and Thomas McCarthy (Cambridge, Mass: The MIT Press, 1987), p. 433.

52. Sinha, p. 352.

53. Ibid., p. 342.

54. Ibid., p. 341.

55. Ibid., p. 340ff.

56. George Gale, "The Anthropic Principle", Scientific American, 245 (June 1981), p. 154-171.

57. Mayr, p. 591.

58. Ibid., p. 589.

59. Ibid., p. 103-105.

60. Ibid., p. 105.

61. Ibid., p. 120.

62. Ibid., p. 570.

63. Ibid., p. 307.

64. Ibid., p. 24.

65. Hilary Putnam, "Why Reason Can't Be Naturalized," in After Philosophy, ed. Kenneth Baynes, James Bohman, and Thomas McCarthy (Cambridge, Mass: The MIT Press, 1987), p. 225.

Reason's Golden Age

1. Timothy Ferris, "Einstein's Wonderful Year", Science 84, 5 (November 1984), p. 61-63.

2. Ibid., p. 62.

3. John Maddox, "How Special is Special Relativity?", Nature, 313 (7 Feb 1985), p. 429.

4. Heinz R. Pagels, The Cosmic Code, (New York: Bantam Books, 1982), p. 36.

5. Orr E. Reynolds, ed., "Proceedings Seventh Annual Meeting IUPS Commission on Gravitational Physiology" Physiologist 28 (December 1985 supplement).

6. Edward Hussey, Aristotle's Physics Books III and IV (Oxford: Clarendon Press, 1983), p. 138-175.

7. Bernard H. Laventa, "Brownian Motion", Scientific American 252 (Feb 1985), p. 70.

8. Pagels, p. 43.

9. Ibid., p. 79-80.

10. Emily Grosholz, Review Article "A New View of Mathematical Knowledge," Brit. J. Phil. Sci., 36 (1985), p. 71.

11. Nancy Cartwright, How the Laws of Physics Lie, (Oxford: University Press, 1983).

12. Pagels, p. 69ff.

13. Ibid., p. 46-48.

14. Ibid., p. 64.

15. Ibid., p. 48.

16. J. Bronowski, The Ascent of Man. (Boston: Little, Brown & Co., 1973), p. 365.

17. Edward Hussey, Aristotle's Physics Books III and IV, (Oxford: Clarendon Press, 1983), p. 188.

18. Pagels, p. 56.

19. Ernst Mayr, The Growth of Biological Thought (Cambridge, Mass: Belknap Press, 1982), p. 24.

20. Pagels, p. 48, p. ll4ff.

21. Ibid., p. 131.

22. Ibid., p. 139-141.

23. Ibid., p. 142.

24. Ibid., p. 157.

25. Ibid., p. 83-84, p. 307.

26. Ibid., p. 187.

27. Ibid., p. 186.

28. Ibid., p. 187.

29. Ibid., p. 283.

30. Ibid., p. 192ff.

31. Ibid., p. 198ff.

32. Ibid., p. 211ff.

33. Ibid., p. 233-234.

34. Ibid., p. 283.

35. Ibid., p. 234.

36. Stuart L. Shapiro, Richard F. Start, and Saul A. Teukolsky, "The search for gravitational waves", American Scientist 73 (May-June 1985), p. 248-257.

37. George Gale, "The Anthropic Principle", Scientific American, 245 (June 1981), p. 154.

38. Bryce S. DeWitt, "Quantum Gravity", Scientific American, 249 (December 1983), p. 112-129. Also Pagels, p. 276.

39. Pagels, p. 259.

40. Ibid., p. 262.

41. Ibid., p. 268.

42. Ibid., p. 304.

43. Ibid., p. 301.

44. Ibid., p. 283.

45. Ibid., p. 307ff.

46. Diogenes Allen, Philosophy for Understanding Theology, (Atlanta: John Knox Press, 1985), p. 24ff.

The Fabric of Theory and Myth

1. Mary Hesse, Revolutions and Reconstructions in the Philosophy of Science (Bloomington: Indiana University Press, 1980), p. 186.

2. Jürgen Habermas, "Philosophy as Stand-In and Interpreter," in After Philosophy, ed. Kenneth Baynes, James Bohman, and Thomas McCarthy (Cambridge, Mass: The MIT Press, 1987), p. 313.

3. Hesse, p. vii-viii, 29ff, 209, 239.

4. Ibid., p. 246.

5. Kenneth Baynes, James Bohman, and Thomas McCarthy, "General Introduction", in After Philosophy, ed. Kenneth Baynes, James Bohman, and Thomas McCarthy (Cambridge, Mass: The MIT Press, 1987), p. 4.

6. Jean-Francdois Lyotard, "The Postmodern Condition," in After Philosophy, ed. Kenneth Baynes, James Bohman, and Thomas McCarthy (Cambridge, Mass: The MIT Press, 1987), p. 80ff.

7. Ibid., p. 86.

8. Hesse, p. 86, p. 131.

9. Ibid., p. 203.

10. Ibid., p. 93.

11. Ibid., p. 126. also p. xii.

12. Ibid., p. xv.

13. Ibid., p. 173.

14. Ibid., p. 63ff.

15. Ibid., p. 184.

16. Ibid., p. 182.

17. Ibid., p. 182-183, also Heinz Pagels, The Cosmic Code (New York: Bantam Books, 1983), p. 114ff.

18. Hesse, p. 125ff.

19. George Gale, "The Anthropic Principle", Scientific American, 245 (June 1981), p. 154-171.

20. Hesse p. 48, 135, 138, 153.

21. Lyotard, p. 81.

22. Richard Rorty, "Pragmatism and Philosophy," in After Philosophy, ed. Kenneth Baynes, James Bohman, and Thomas McCarthy (Cambridge, Mass: The MIT Press, 1987), p. 31.

23. Hesse, p. 159.

24. Ibid., p. 174.

25. Ibid., p. 158.

26. Ibid., p. 167.

27. Ibid., p. 175.

28. Baynes et al, p. 16.

29. Alasdair MacIntyre, "Relativism, Power, and Philosophy," in After Philosophy, ed. Kenneth Baynes, James Bohman, and Thomas McCarthy (Cambridge, Mass: The MIT Press, 1987), p. 393-394.

30. Hesse, p. 187ff.

31. Ibid., p. 246.

32. Ibid., p. 208ff.

33. Lyotard, p. 89.

34. Hesse, p. 217.

35. Lyotard, p. 89.

36. Hilary Putnam, "Why Reason Can't Be Naturalized," in After Philosophy, ed. Kenneth Baynes, James Bohman, and Thomas McCarthy (Cambridge, Mass: The MIT Press, 1987), p. 229.

37. Charles Downey, "Is Man Alone in the Universe? Yes, Scientists Insist." Sacramento Bee, March 3, 1985, p. G6.

38. J. Richard Gott, III, James E. Gunn, David N. Schramm and Beatrice M. Tinsley, "Will the Universe Expand Forever?", in Cosmology + 1, ed. Owen Gingerich, (San Francisco: W. H. Freeman & Co., 1977), p. 82-93.

Science: Promise of the Covenant

1. Mortimer J. Adler, Ten Philosophical Mistakes, (New York: McMillan Publishing Co., 1985), p. 108ff.

2. Tom D'Evelyn, "Cohen's Love Song to Science: A History of Revolutions," The Christian Science Monitor, April 24, 1985, p. 21. (on a review of I. Bernard Cohen, Revolution in Science (Cambridge, Mass: Belknap Press, 1985), p. 147.

3. John Marks, Science and the Making of the Modern World, (London: Exeter, 1983), p. 490.

4. Allen L. Hammond ed., "20 Discoveries That Changed Our Lives", Science 84, 5 (November 1984).

5. Ibid., p. 9.

6. Allen L. Hammond, ed., "The Next Step: 25 Discoveries that Could Change Our Lives", Science 85, 6 (November 1985).

7. Kenneth J. Silverman and William Grossman, "Current Concepts Angina Pectoris", New England Journal of Medicine, 310 (June 28, 1984), p. 1712-1717.

8. Lee Goldman and E. Francis Cok, "Reviews, The Decline in Ischemic Heart Disease Mortality Rates," Annals of Internal Medicine, 101 (December 1984), p. 825-836.

9. Slavka Frankova, "Lasting Effects of Early Malnutrition on Children's Behavior", in Nutrition in Early Childhood and its Effects in Later Life, ed. J. C. Somogyi and H Haenel (Basel: S. Karger, 1982), p. 40-54.

10. Nicholas Maxwell, From Knowledge to Wisdom, (New York: Basil Blackwell, 1984), p. 55.

11. M. Mitchell Waldrop, "First Sightings", Science 85, 6 (June 1985), p. 26.

12. George Greenstein, "An Invitation to Strangers", Science 85, 6 (June 1985), p. 41.

13. Ibid., p. 40.

14. Kip S. Thorne, "The Search for Black Holes", in Cosmology + 1, ed. Owen Gingerich (San Francisco: W. H. Freeman & Co.. 1977), p. 69.

15. Heinz R. Pagels, The Cosmic Code (Toronto: Bantam Books, 1982), p. 286-287.

16. Robert C. Cowen, "Is High-Energy Physics Worth its High Cost?" The Christian Science Monitor, March 26, 1985, pp. 25, 27.

17. R. N. Hirons, "Mathematical Models in Demography," in The Structure of Human Populations, ed. G. A. Harrison and A. J. Boyce, (Oxford: Clarendon Press, 1972), p. 112ff.

18. Ibid., p. 114.

19. Barbara Burke, "Infanticide", Science 84, 5 (May 1984), p. 26-31.

20. Burton Benedict, "Social Regulation of Fertility," in The Structure of Human Populations, ed. G. A. Harrison and A. J. Boyce, (Oxford: Clarendon Press, 1972), p. 73-89.

21. Lee-Jay Cho, Wilson H. Grabill and Donald J. Bogue, Differential Current Fertility in the United States (Chicago: Community and Family Study Center, University of Chicago, 1970).

22. Werner Fornos, "Toward Voluntary Poppulation Control", The Christian Science Monitor, August 7, 1984, p. 15.

23. David K. Willis, "Locating the 'Third World,'" The Christian Science Monitor, November 4, 1985, p. 25.

24. Colin McEvedy and Richard Jones, Atlas of World Population History, (London: A. Lane, 1978), p. 343.

25. Ibid., p. 345.

26. Ibid., p. 345-346.

27. Ibid., p. 346ff.

28. David K. Willis, "Africa's Urgent Need," The Christian Science Monitor, November 27, 1984, p. 24-25.

29. David K. Willis, "A Tidal Wave of Humanity," The Christian Science Monitor, August 6, 1984, p. 12-13.

30. David K. Willis, "Seeds of Revolution," The Christian Science Monitor, November 28, 1984, p. 22-23.

31. David K. Willis, "Overcrowding: The Impact and the Risks," The Christian Science Monitor, August 8, 1984, p. 20-21.

32. David K. Willis, "The Supercities," The Christian Science Monitor, August 7, 1984, p. 18-19.

33. W. M. Henderson, "Identification of Existing and Prospective Problems of Disease Control" in Virus Diseases of Food Animals vol. I, ed. E. P. J. Gibbs, (New York: Academic Press, 1981), p. 14.

34. E. P. J. Gibbs, ed., Virus Diseases of Food Animals 2 vol. (New York; Academic Press, 1981).

35. Henderson, p. 14-15.

36. Ibid., p. 14.

Humanizing the Species

1. Werner Stark, The Social Bond, Vol. I. (New York: Fordham University Press, 1976), p. 177ff.

2. Ibid., p. 177.

3. Ibid., p. 179ff.

4. Ibid., p. 183.

5. Ibid., p. 178.

6. Ibid., p. 180ff.

7. Ibid.

8. Ibid., p. 185, quote of Emile Durkheim.

9. Ibid., p. 122ff.

10. Ibid., p. 216, quote of Edward O. Wilson.

11. Ibid., p. 215ff.

12. Wn Rukang and Lin Shenglong, "Peking Man", Scientific American, 248 (June 1983), p. 94.

13. "Currents" Science 86, (March 1986), p. 6.

14. Stark, p. 179.

15. Frank Rose, "The Black Knight of AI," Science 85, 6 (March 1985), p. 47-48.

16. M. Mitchell Waldrop, "Machinations of Thought," Science 85, 6 (March 1985), p. 45.

17. Jerome Kagan, "On Love and Violence," Science 85, 6 (March 1985), p. 29.

18. Mary Hesse, Revolutions and Reconstructions in the philosophy of Science (Bloomington: Indiana University Press, 1980), p. 187ff.

19. Joseph Alper, "The Roots of Morality," Science 85, 6 (March 1985), p. 74.

20. Jacques Derrida, "The Ends of Man" in After Philosophy, ed. Kenneth Baynes, James Bohman, and Thomas McCarthy (Cambridge, Mass: The MIT Press, 1987), p. 89.

21. Stark, p. 6ff.

22. Ibid., p. 103-104.

23. Ibid., p. 103ff.

24. Ibid., p. 107.

25. Ibid., p. 108ff.

26. Mortimer J. Adler, Ten Philosophical Mistakes, (New York: McMillan Publishing Co., 1985), p. 191ff.

27. Stark, p. 133.

28. Ibid., p. 134-135.

29. Ibid., p. 134.

30. Ibid., p. 138.

31. Ibid., p. 150ff.

32. Ibid., p. 213.

33. Ibid., p. 149-150 quote of Imre Madach.

34. F. A. Hayak, Law, Legislation and Liberty, vol 1 Rules and Order (London: Routledge & Kegan Paul, 1973), p. 46.

35. Richard Rorty, "Pragmatism and Philosophy," in After Philosophy, ed. Kenneth Baynes, James Bohman, and Thomas McCarthy (Cambridge, Mass: The MIT Press, 1987), p. 89.

36. Stark, p. 189ff.

37. Ernst Mayr, The Growth of Biological Thought (Cambridge, Mass: Belknap Press, 1982), p. 622.

The Purpose of Science

1. Alexander Keller, "Has Science Created Technology?", Minerva, 22 (Summer 1984), p. 172, quoting F. R. Jevons.

2. Kenneth Scott Latourette. A History of Christianity, (New York: Harper and Brothers, 1953), p. 260.

3. Ernst Mayr, The Growth of Biological Thought (Cambridge, Mass: Belknap Press, 1982), p. 38.

4. Mortimer J. Adler. Ten Philosophical Mistakes, (New York: McMillan Publishing Co., 1985), 191ff.

5. Mary Hesse, Revolution and Reconstruction in the Philosophy of Science (Bloomington: Indiana University Press, 1980), p. 248-249.

6. J. Bronowski, The Ascent of Man. (Boston: Little, Brown & Co., 1973), p. 364.

7. Bertrand Russell, Wisdom of the West (Garden City, N.Y.: Doubleday & Co., 1959), p. 274-275.

8. The Holy Bible, Gen. 8:21.

9. Hans-Georg Gadamer, "The Transformation of Philosophy: Hermeneutics, Rhetoric, Narrative", in After Philosophy, ed. Kenneth Baynes, James Bohman, and Thomas McCarthy (Cambridge, Mass: The MIT Press, 1987).

In the Beginning There Was Value

1. Nicholas Maxwell, From Knowledge to Wisdom (New York: Basil Blackwell, 1984), p. 51-52.

2. Ibid., p. 47ff.

3. Ibid., p. 52.

4. Ibid., p. 59-61.

5. The Holy Bible, Judges 17:6.

6. Stephen Toulmin, The Return to Cosmology (Berkeley: University ofCalifornia Press, 1982), p. 59, summarizing Julian Huxley.

7. Ibid., 144.

8. Maxwell, p. 79. 100.

9. Diogenes Allen, Philosophy for Understanding Theology (Atlanta: John Knox Press, 1985), p. 134.

10. The Holy Bible, Job 28:12-14, 21, 23.

11. Ibid., Job 26:14.

12. Ibid., Joshua 24:15.

13. Ibid., Psalms 91:1-2.

14. Ibid., Joshua 24:15.

INDEX